JOURNAL OF CYBER SECURITY AND MOBILITY

Volume 1, No. 4 (October 2012)

Special issue on

Next Generation Roaming Applications: Implementation Challenges for Rich Communication Suite Applications, Locations & Presence Based Services and NFC for Roaming Scenarios in Heterogeneous Networks

Guest Editors:

Rajarshi Sanyal[1] and Pascal Alloin[2]
[1]*Engineering and Operations Expert, Belgacom International Carrier Services*
[2]*Director of Engineering, Belgacom International Carrier Services*

JOURNAL OF CYBER SECURITY AND MOBILITY

Editors-in-Chief
Ashutosh Dutta, AT&T, USA
Ruby Lee, Princeton University, USA
Neeli R. Prasad, CTIF-USA, Aalborg University, Denmark

Associate Editor
Shweta Jain, York College CUNY, USA

Steering Board
H. Vincent Poor, Princeton University, USA
Ramjee Prasad, CTIF, Aalborg University, Denmark
Parag Pruthi, NIKSUN, USA

Advisors
R. Chandramouli, Stevens Institute of Technology, USA
Anand R. Prasad, NEC, Japan
Frank Reichert, Faculty of Engineering & Science University of Agder, Norway
Vimal Solanki, Corporate Strategy & Intel Office, McAfee, Inc, USA

Editorial Board

Sateesh Addepalli, CISCO Systems, USA
Mahbubul Alam, CISCO Systems, USA
Jiang Bian, University of Arkansas for Medical Sciences, USA
Tsunehiko Chiba, Nokia Siemens Networks, Japan
Debabrata Das, IIIT Bangalore, India
Subir Das, Telcordia ATS, USA
Tassos Dimitriou, Athens Institute of Technology, Greece
Pramod Jamkhedkar, Princeton, USA
Eduard Jorswieck, Dresden University of Technology, Germany
LingFei Lai, University of Arkansas at Little Rock, USA
Yingbin Liang, Syracuse University, USA
Fuchun J. Lin, Telcordia, USA

Rafa Marin Lopez, University of Murcia, Spain
Seshadri Mohan, University of Arkansas at Little Rock, USA
Rasmus Hjorth Nielsen, Aalborg University, Denmark
Yoshihiro Ohba, Toshiba, Japan
Rajarshi Sanyal, Belgacom, Belgium
Andreas U. Schmidt, Novalyst, Germany
Remzi Seker, University of Arkansas at Little Rock, USA
K.P. Subbalakshmi, Stevens Institute of Technology, USA
Reza Tadayoni, Aalborg University, Denmark
Wei Wei, Xi'an University of Technology, China
Hidetoshi Yokota, KDDI Labs, USA

Aim

Journal of Cyber Security and Mobility provides an in-depth and holistic view of security and solutions from practical to theoretical aspects. It covers topics that are equally valuable for practitioners as well as those new in the field.

Scope

The journal covers security issues in cyber space and solutions thereof. As cyber space has moved towards the wireless/mobile world, issues in wireless/mobile communications will also be published. The publication will take a holistic view. Some example topics are: security in mobile networks, security and mobility optimization, cyber security, cloud security, Internet of Things (IoT) and machine-to-machine technologies.

JOURNAL OF CYBER SECURITY AND MOBILITY

Volume 1   No. 4   October 2012

Published, sold and distributed by:
River Publishers
P.O. Box 1657
Algade 42
9000 Aalborg
Denmark

Tel.: +45369953197
www.riverpublishers.com

Journal of Cyber Security and Mobility is published four times a year.
Publication programme, 2012: Volume 1 (4 issues)

ISSN 2245-1439 (Print Version)
ISSN 2245-4578 (Online Version)
ISBN 978-87-92982-44-5 (this issue)

Editorial Foreword

Roaming is an important aspect in mobile communication that renders the ability to operate the mobile device beyond the fences. A few years back, Voice, SMS and GPRS were the primal drivers to facilitate roaming. But with the advent of the smartphones, the recent trend is to render the Value Added Services and the applications with the same look and feel in all roaming networks as being at home. This implies that the roaming brokers should not only act as carriers for providing the basic services (Voice/SMS/GPRS), but they should also be able to cater the plethora of VAS and Rich Communication Suite (RCS) applications which are the game changers. The evolution of the state of the art mobile network is propelled by the requirements of the next generation users where interoperability, security, service continuity, seamless operations in ubiquitous network domains and policy enforcement are some of the important factors. To cater these demands, especially in light of the roaming paradigm, the service providers face some challenges:

- Guarantee bandwidth for the mobile user
- Allow the mobile terminal to operate ubiquitously across various access methods (LTE/GSM/WiFi/CDMA/WIMAX) and to manage the authentication and mobility at the core network layer
- Simplification of roaming access and management

When a mobile user initiates a data session from the mobile terminal, the visited network attempts to establish a packet data context with the gateway at the home network connected to internet. Subsequently, the packet data is conveyed between the mobile terminal and the data content provider. The packet data is logically channelized via the visited network-home network route, using conventional mobile data exchange protocols like GTP (GPRS Tunnelling Protocol) or the MIP (Mobile IP). Routing to home network can however prove to be expensive, which is why the operators are investigating solutions to offload traffic at the visited network itself. This is commonly referred to as 'Mobile Data Offload'. So in case of a LTE network, the GTP traffic which is routed from the Serving Gateway of the visited network will be typically offloaded locally at the Packet Gateway of the visited network over the S5 interface, instead of the Packet Gateway through the S8 interface. This is termed as Local Breakout. However, the policy control and charging guidelines usually stay at the home network. This is why the visited network may need to interact with the PCRF (Policy and Charging Rules Function) through the S9 interface. The basic authentication and mobility related parameters are also governed by the home network. More specifically, the Mobility Management Entity (MME) in the LTE core of the visited network exchanges these parameters over S6 interface with HSS (Home Subscriber Server) to actuate authentication and enable creation of a dynamic profile of the roamer in the visited network. Both S6 and S9 interfaces follow the Diameter protocol which has been enriched to support the LTE authentication and mobility management functions by coining new Diameter AVPs (Attribute Value Pairs). As S6 and S9 interfaces between the visited and the core network is only related to signalling, it is relatively lightweight in terms of bandwidth, when compared to the actual data traffic which can be locally offloaded. Local Breakout is patronised by European Commission through the organisation called BEREC (Body of European Regulators for Electronic Communications).

With the aim of further network optimisation, some operators are considering a local break out of the Data traffic towards the internet cloud through satellite backhaul at the access network layer. A network element that can emulate the GSN (GPRS Support Node) components can be implemented at the BSC (Base Station Controller) to translate the GTP data traffic to the TCP/IP traffic for routing it to the content provider.

Cross technology mobility management is another key topic. As the smartphones support various access technologies and frequency bands, the big question is how to support inter-domain mobility management and bring out new services and revenue streams. A point in case is WiFi roaming. When we access WiFi within the home or roaming location, we have to opt for a WiFi subscription or may have to search for a free WiFi network. Picking a WiFi subscription while you are in transit may prove to be difficult. The free WiFi Access in the roaming location may not provide you the necessary bandwidth and reliability that you require for your application, for example video conferencing.

Telecom researchers are finding out ways to reuse the 3G authentication mechanism to actuate AAA (Authentication, Authorisation and Accounting) in the WiFi domain. So the user can use the existing GSM subscription to acquire a WiFi network in the roaming domain. A central platform aggregates the AAA information from all the WiFi providers and routes this signalling traffic over RADIUS protocol to an application platform. This platform translates RADIUS to GSM MAP (Mobile Application Part) and forwards to the GSM domain. The authentication and authorisation is done by the core network following which acknowledgement message is invoked and routed back to the platform. The various protocols relevant are WISPr, EAP-AKA/SIM.

In the good old days, roaming agreements were established through bilateral means. As the number of networks increased, augmenting and maintaining the roaming relations proved to be difficult and expensive. Hence came in the concept of an OC-RH (Open Connectivity Roaming Hub) coined by GSMA. Typically a Roaming Hub will enable to build up a roaming consortium via direct connectivity with the mobile networks, or through peering agreements with other Roaming Hubs. So when a mobile operator connects to this platform, it gets immediate roaming access to the plethora of mobile networks. The primary functions like mobility management, SMS Interworking, Roaming Billing Settlement are catered by an OC-RH.

Network Functions Virtualisation over the cloud is a new concept which some operators are following closely. The state of the art mobile network comprises of core network elements from multiple vendors implemented at the operator premises, implying substantial CAPEX (Capital Expenditure) and OPEX (Operational Expenditure). Hence some leading operators in USA, Europe and Far East jointly made a proposition to implement the Mobile Core Network Elements on the cloud for optimisation, consolidation, cost reduction and boosting operational efficiency. The primary concerns are however security, reliability , latency and mobility/roaming.

In spite of all the technological advancements in the roaming landscape, the conundrum still remains: How do we realise a network, intelligent enough to ensure harmonized service delivery across the heterogeneous topologies?

To solve the many elements of this jigsaw puzzle, we first address the issue of interworking. As the modern mobile users, be it human or machine, become progressively dynamic in the network space, most of the terminals more frequently need to interoperate and switch between diverse access technologies. This will be meaningless unless it is supported by a robust interworking framework at the core network. The first paper from Arnab Dey et al. (Diametriq) attempts to carve a feasible model to render interworking between the LTE and the legacy networks. Bridging the mobility management procedures across the 2 network domains is an eternal problem which is eloquently addressed in this paper.

The second paper from Parikshit Mahalle et al. dwells on security and policy control for Internet of Things (IoTs) where mobility plays an important function. The role of authentication and access control in contemporary research is brought out and the bottlenecks are identified. A new model, IACAC is proposed in light of the 'state of the art' network paradigms namely LTE , Wifi, NFC.

Mobile number portability allows a customer to retain the ISDN address during migration from one operator to other. Number continuity is another important aspect which enables a mobile user to switch domains transparently retaining the home number. Arnaud Henry-Labordère's (Halys) paper on number continuity between GSM and satellite, portrays the practical aspects of such an implementation. It delves in the mobility management and handover issues and sets out the architecture for delivering the tele-services and supplementary services ubiquitously during this transition.

Cellular roaming implies that that the home network engages the visited network not only in the mobility management processes, but also during authentication and authorization of users. We live in a complex world where identity theft in telecom networks is common, and the mobile networks are not exceptions. The fourth paper from Geir Køien draws parallelism between global roaming and Virtual Machines (VM) on cloud and proposes a spatio-temporal exposure model to mitigate the risks related to mobility, migration and identity piracy.

We have some high quality research papers in this issue notched out from academics and industry. The papers attempt to bring out the theoretical approaches as well as the practical aspects of the problems encountered during the perpetual process of network evolution. We anticipate that this issue will contribute positively by shaping new ideas for the vanguards of this evolution.

Rajarshi Sanyal and Pascal Alloin

Realization of Interworking in LTE Roaming Using a Diameter-based Interworking Function

Arnab Dey, Balaji Rajappa and Lakshman Bana

Engineering, Diametriq, Melbourne, USA;
e-mail: {adey, brajappa, lbana}@diametriq.com

Received 15 January 2013; Accepted 17 February 2013

Abstract

While operators around the world are onboarding or planning to onboard the LTE-based Evolved Packet System (EPS), some troubling questions linger on. The prospects of LTE are enticing. However, would the LTE and Legacy networks interwork and co-exist? Would inbound/outbound roamers face any service disruption? For a green-field operator, possibly with a tight budget, is adopting LTE a risky proposition when over 90% of the world's network is still based on Legacy Signaling System No. 7 (SS7)? Various standards bodies have provided guidelines and specifications to identify and address some of the interworking and co-existence scenarios, but their implementation is complex and requires a detailed knowledge of the disparate worlds of EPS and Legacy protocols in finding an acceptable intersection between the two.

This paper describes the important interworking issues between Diameter and TCAP (SS7)-based protocols and some of the practical aspects that transcend information that is disseminated through the standards bodies. Some specific scenarios such as a 2G/3G subscriber roaming into an LTE network using home-routed applications and S8 (EPS) – Gp (Legacy) interworking are not covered here and shall be discussed in a subsequent paper. Similarly, this paper has not focused on Diameter–RADIUS interworking needed to support RADIUS-based AAA Server. However, most other scenarios have

Journal of Cyber Security and Mobility, Vol. 1, 289–308.

been addressed and it is hoped that the solution presented here will alleviate the roaming, co-existence and interworking concerns of the reader.

Keywords: IWF, interworking, diameter, TCAP, LTE, MAP, CAP.

Abbreviations

AAA	Authentication Authorization Accounting
ACN	Application Context Name
ASN.1	Abstract Syntax Notation 1
AVP	Attribute Value Pair
CAP	CAMEL Application Part
EIR	Equipment Identity Register
EPS	Evolved Packet System
E-UTRAN	Evolved UTRAN
GERAN	GPRS Edge RAN
GPRS	General Packet Radio Service
HLR	Home Location Register
HPLMN	Home PLMN
IE	Information Element
IMSI	International Mobile Subscriber Identity
IWF	Inter Working Function
LTE	Long Term Evolution
MAP	Mobile Application Part
MCC	Mobile Country Code
MNC	Mobile National Code
MSIN	Mobile Subscriber Identification Number
OCS	Online Charging System
PCC	Policy and Charging Control
PLMN	Public Land Mobile Network
QoE	Quality of Experience
RAN	Radio Access Network
RADIUS	Remote Authentication Dial-In User Svc
SCP	Service Control Point
SCTP	Stream Control Transmission Protocol
SIGTRAN	Signaling Transport
SS7	Signaling System No. 7
TBCD	Telephony Binary Coded Decimal
TCAP	Transaction Capabilities Application Part

TCP	Transmission Control Protocol
UE	User Equipment
UTRAN	Universal Terrestrial RAN
VPLMN	Visited PLMN

1 Introduction

1.1 Understanding of Common Terms

The terms LTE, E-UTRAN, EPC and EPS have been used at various places in the text. It is important to have a lucid understanding of these terms and how they relate to each other before proceeding further.

The terms LTE and E-UTRAN are commonly used interchangeably. E-UTRAN actually refers to the RAN that uses LTE, which is the radio interface technology.

EPC is the core network that uses E-UTRAN.

EPS comprises the UE, E-UTRAN, EPC and other access networks connecting through the EPC.

LTE also being the name of the 3GPP work item that developed E-UTRAN and the corresponding radio interface technology is normally used in a broader scope in everyday usage.

1.2 The Lure of LTE

The LTE radio network provides an enriched Quality of (User) Experience (QoE) through higher peak data rates, lower latency and higher spectral efficiency. Thus, it can support the needs of today's bandwidth hogging Internet applications. The EPS architecture is designed to interwork 2G/3G, trusted non-3GPP (such as CDMA) and non-trusted non-3GPP (such as WLAN) accesses. The centralized Policy Control and Charging (PCC) framework in EPS allows Subscriber and Service differentiation, at the same time paving the way for optimized user and network resource usage.

1.3 The Dilemma with LTE

Had EPS been a monolithic architecture comprising only LTE, E-UTRAN and EPC, the promise of EPS' "advanced" technology would have been more palatable to the user community. However, EPS is supposed to interwork with heterogeneous networks, most of which have been there for a long time. Each

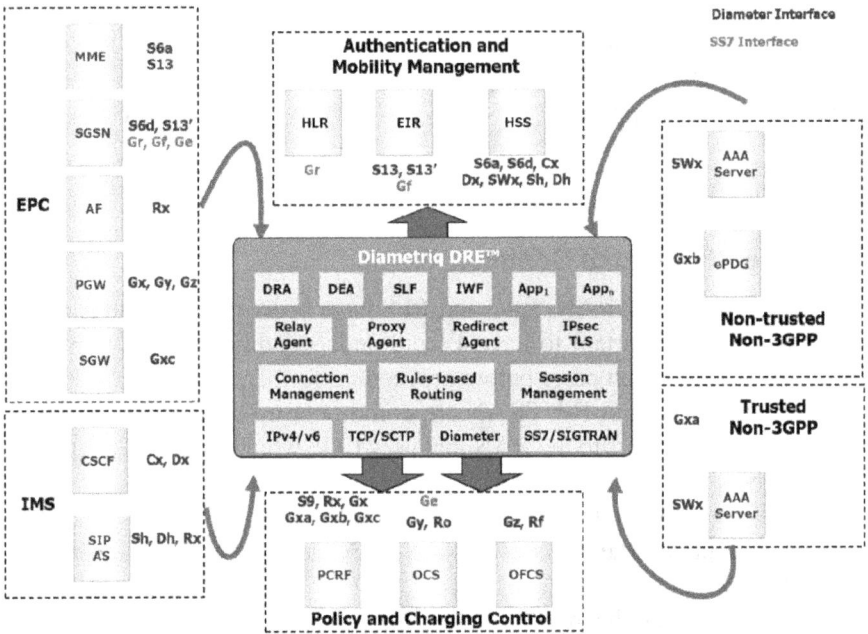

Figure 1 Diameter & SS7 interfaces in the EPS ecosystem.

of these networks has its unique technology foundation. Integrating the varied technologies into the single EPS umbrella is a daunting task.

1.4 Towards a Solution

Appreciating the magnitude of the task, standards bodies have started specifying the various and various kinds of interworking scenarios. The approach has been to first identify and address those scenarios that roamers are more likely to encounter.

"Diameter" is the protocol for Authentication, Authorization and Accounting (AAA) in the EPS. It is used for Authentication, Mobility Management, Policy and Charging Control procedures. The Diameter interfaces supporting these procedures are illustrated in Figure 1. Furthermore, Table 1 illustrates the various SS7 interfaces and the corresponding serving network entities.

This paper focuses on the interaction of the Gr, Gf and Ge SS7 and related Diameter interfaces in the following scenarios:

Table 1 Mapping of SS7 interfaces into serving network entities.

SS7 Interface	Network Entity
Gr	HLR
Gf	EIR
Ge	OCS (Prepaid SCP)

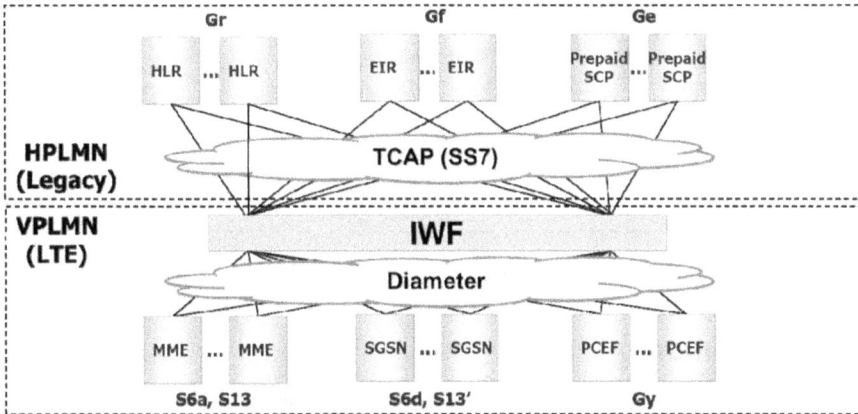

Figure 2 HPLMN is Legacy, VPLMN is LTE.

- HPLMN is Legacy, VPLMN is LTE – A 2G/3G subscriber roaming into an E-UTRAN.
- HPLMN is LTE, VPLMN is Legacy – A 4G subscriber roaming into an GERAN/UTRAN.

The above roaming scenarios are pictorially depicted in Figures 2 and 3. The InterWorking Function (IWF) mediates between two disparate protocols and makes them interoperable. It is assumed that the IWF is located in the EPS.

Table 2 Diameter – TCAP interworking in various roaming scenarios.

HPLMN is Legacy, VPLMN is LTE		HPLMN is LTE, VPLMN is Legacy	
Diameter	TCAP	TCAP	Diameter
S6a	Gr (MAP)	Gr (MAP)	S6d
S6d	Gr (MAP)	Gf (MAP)	S13'
S13	Gf (MAP)	Ge (CAP)	Gy
S13'	Gf (MAP)		
Gy	Ge (CAP)		

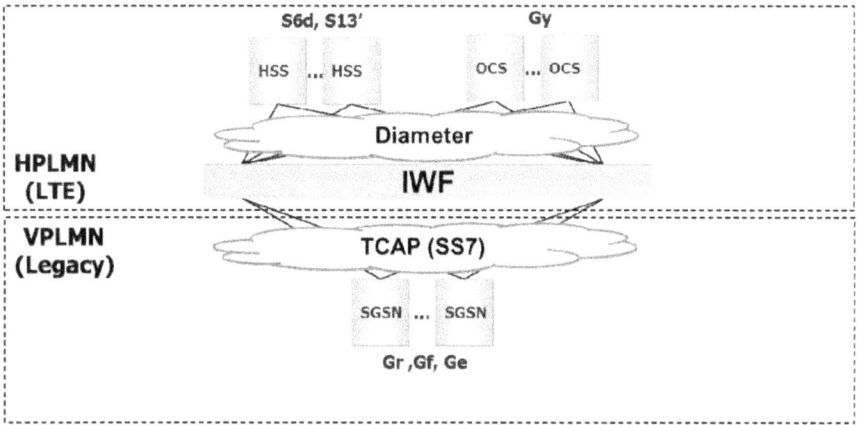

Figure 3 VPLMN is Legacy, HPLMN is LTE.

2 The Solution

The upcoming subsections unravel the nuts and bolts of the IWF.

2.1 IWF Stack Diagram

The IWF comprises a layered architecture having the following layers:

- Protocol – Every transaction in an interworking scenario involves two underlying protocols, for example, Diameter and TCAP. These operate independent of each other and comply with their "separate" protocol specifications as far as parameter and message encoding and decoding, message routing and protocol management is concerned. Supporting a new interworking entails supporting a new protocol.
- Transport Handler – This abstracts the usage of the corresponding underlying transport from the higher layers. Thus, there is one Transport Handler relevant to each protocol. Diameter is supported over TCP/SCTP. TCAP is supported over SIGTRAN/SS7.
- Transaction Manager – This is a singleton that manages the interworking transactions/sessions. It uses a State Machine in determining how to deal with an incoming request or response, and whether protocol conversion should be invoked. The beauty of the State Machine is its generality, which lends easy extensibility to the IWF in supporting a newer interworking.

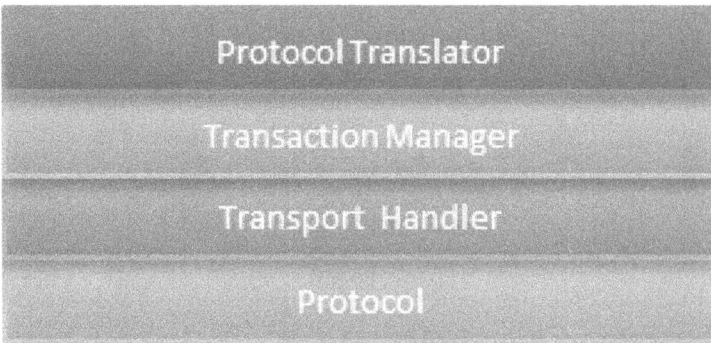

Figure 4 IWF state diagram.

- Protocol Translator – This is what performs the parameter and message mapping from one protocol to the other. Some of the protocol translations have been defined by standards bodies, for example, that between Diameter and MAP based interfaces [1]. Some others, for example, that between Diameter and CAP, have not yet been defined by any standards body – such protocol translations can be defined through knowledge of the corresponding protocols and interworking scenarios.

The IWF layered architecture has been depicted in Figure 4.

2.2 Transport Handling

Transport handling depends on the idiosyncrasies of a particular protocol. While a discussion of a protocol involved in an interworking and its transport related aspects is outside the scope of this paper, it is worthwhile to represent the two protocol suites used as baseline in this paper (Diameter & TCAP). This helps us get a peek into the disparate nature of the two protocols and the ensuing complexity of interworking – an interested but uninitiated reader is encouraged to study the relevant protocol specifications to appreciate this even better.

2.3 Transaction Manager

This is the *control center* of the IWF. It is invoked when a protocol message arrives. At the time of invocation it knows the protocol of the received message. It then determines the following:

1. Whether the message is a request or response.

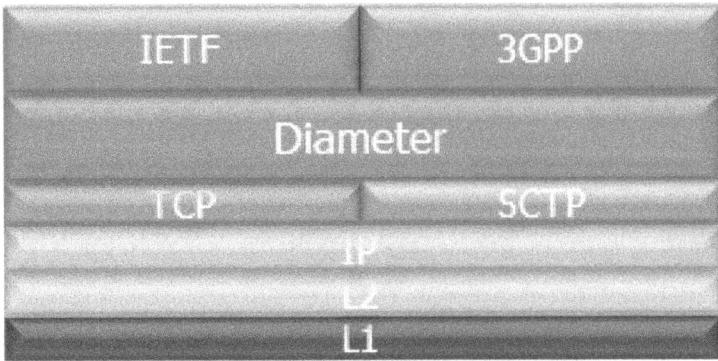

Figure 5 Diameter protocol suite.

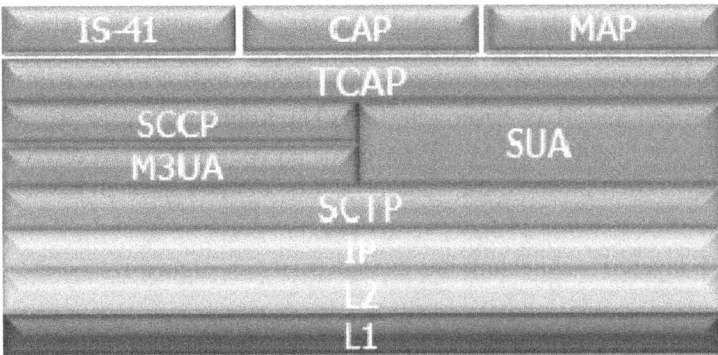

Figure 6 SIGTRAN protocol suite.

2. In the case of the TCAP protocol, whether the message is a TCAP dialog or component and what is the dialog or component type. This is because the TCAP protocol is complex in nature. A TCAP request consists of an optional dialog portion and one or more components. Some TCAP primitives are only generated and sent "locally" from the TCAP stack to the Transaction Manager, for example, when a response has not been received before a timer expiry. Refer to [3] for a functional description of the transaction capabilities of TCAP.

3. In the case of the TCAP protocol, the TCAP transaction the TCAP dialog or component belongs to.

4. If it is the "first" message of a transaction, the *protocol at request origination*. For example, when a 2G/3G subscriber roams into an LTE

area and switches on his UE, the ensuing authentication and mobility management procedures cause Diameter S6a/S6d transactions to be initiated at the MME/SGSN towards the HLR (3GPP Release 8 or later). In this case, the protocol at request origination is *Diameter*. Similarly, when in another scenario, an LTE subscriber roams into a 2G/3G area and switches on his UE, the ensuing authentication and mobility management procedures cause MAP Gr transactions to be initiated at the SGSN (3GPP Release 8 or later). In this case, the protocol at request origination is *MAP*.

Based on the message received, the Transaction Manager determines the unique context the message belongs to. In these cases, "all" protocol messages belonging to the "same" transaction contain the "same unique" transaction identifier. In Diameter, this transaction identifier is referred as session identifier (refer to "Session-Id" in [2]). In TCAP, each of the two TCAP end-user applications, for example, the TCAP end-user applications residing in IWF and HLR respectively, is called a TC-user ("TC" refers to Transaction Capabilities). In each transaction in TCAP, each TC-user maintains its own transaction identifier. The origination and destination transaction identifiers are swapped depending on the direction of a message in the same transaction. Unique "dialogue identifier" maps onto the transaction identifiers exchanged in messages belonging to the same transaction in either direction (refer to "Dialogue ID" in [3]). The Transaction Manager maintains the following two mappings for an IWF transaction:

- In the case of a message received on Diameter, a mapping of the Session-Id to a unique IWF context.
- In case of a message received on TCAP, a mapping of the Dialogue ID to the same IWF context.

As a result of the determination of (1), (2), (3), and (4), the Transaction Manager is able to compute the IWF *Event* that has occurred on a transaction. In case of the "first" message of a new transaction it creates a new IWF context and establishes a mapping of the newly created context with the received Diameter Session-Id or TCAP Dialogue ID; else it derives the stored IWF context based on the received Session-Id or Dialogue ID.

On the IWF context, the Transaction Manager then invokes the IWF State Machine, passing it the event and protocol message received. The State Machine has been developed keeping the various interworking scenarios in mind. It is designed to support "any" interworking, remaining unaware of the underlying protocols themselves. Thus the framework can be easily extended

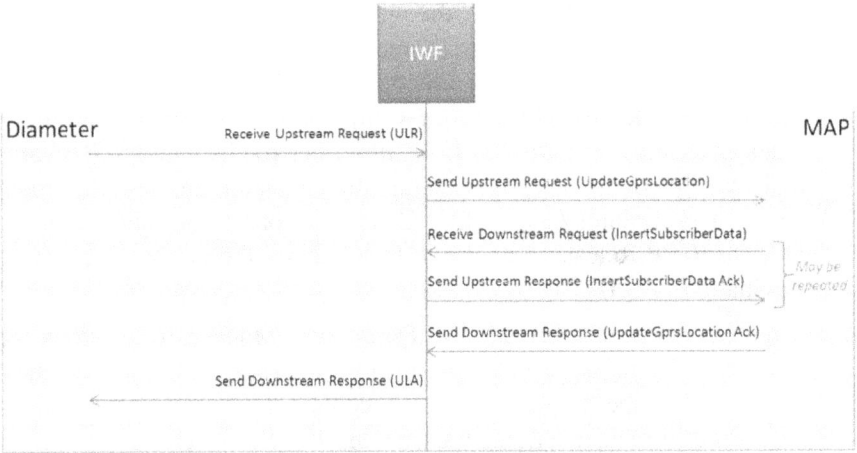

Figure 7 "With" interim request-response on MAP.

to support an additional interworking, for example, Diameter-RADIUS. The State Machine maps a possible incoming event and current state of the IWF context to a succeeding state and corresponding actions.

Figures 7, 8, 9 and 10 depict four examples, which among them represent "all" interworking scenarios for Diameter-MAP shown in Figures 2 and 3. These figures illustrate the generality of the State Machine.

In these figures, "*upstream*" refers to a direction from the Originating/IWF to the Terminating entity, with respect to a request or response. In Diameter-MAP interworking, the Originating entity is MME/SGSN and Terminating entity is HLR; an upstream request originates from MME/SGSN to IWF, or IWF to HLR; an upstream response originates from IWF to HLR. Correspondingly, "downstream" refers to a direction from the Terminating/IWF to the Originating entity, with respect to a request or response. In Diameter-MAP interworking, a downstream request originates from HLR to IWF, or IWF to MME/SGSN; a downstream response originates from HLR to IWF, or IWF to MME/SGSN.

2.4 Protocol Translator

This is the *knowledge engine* of the IWF. It is invoked when the Transaction Manager determines a protocol conversion is required. Implementing protocol translation between disparate protocols requires knowledge and ex-

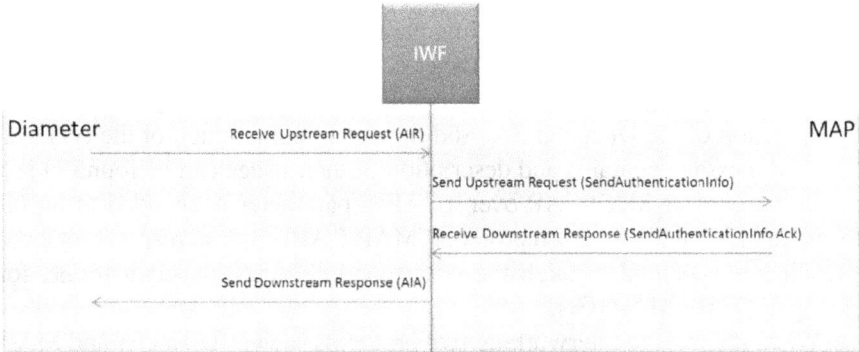

Figure 8 With "no" interim request-response on MAP.

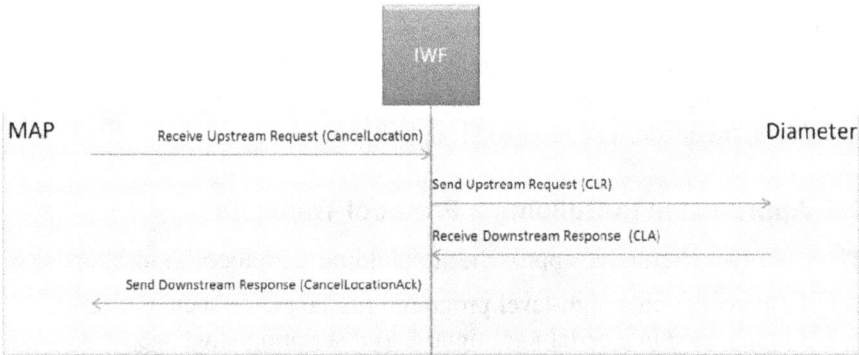

Figure 9 Request originated from MAP (including server-initiated).

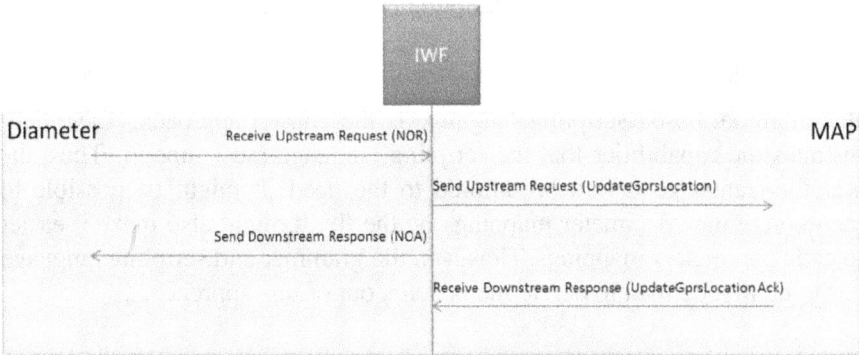

Figure 10 Diameter response sent without waiting for a MAP response.

perience in the encoding and decoding of parameters and an understanding of the correlation between corresponding parameters of the respective protocols.

In the case of Diameter, a parameter is encoded in an Attribute Value Pair (AVP). The concept of AVP and the various AVP types are specified in [2]. Composition of the Diameter S6a/S6d messages, connotation of the AVPs an S6a/S6d message contains and description of their usage can be found in [8].

In the case of MAP/CAP over TCAP, a parameter is encoded in an Information Element (IE). Each protocol (MAP/CAP) request, response or error encoding is specified through the corresponding ASN.1 specifications, for example, [7] for MAP (Gr).

When it comes to interworking two protocols such as Diameter and MAP, the mapping of corresponding parameter is described in a home-grown specification or that from a standards organization. For example, the mapping of Diameter-MAP in Figure 2 is specified in [1]. How to implement the mapping requires algorithmic and programmatic skills.

In Table 2, AVP refers to a field in Diameter and IE refers to a field in MAP (or CAP). For a description of the various field names and types cited in Table 3, the reader can refer to [1, 2, 6–8].

2.5 Approaches to Building a Protocol Translator

There are two alternative approaches to building a Protocol Translator:

1. Logic coded in a high-level programming language such as C/C++.
2. Logic coded in a certain grammar-based scripting language.

Factors determining the chosen approach:

a. Availability of skilled personnel in the language of choice
b. Whether a grammar-based scripting language is already available, or has to be created
c. Time-to-market needs

If a grammar-based scripting language is the chosen approach, Table 3 illustrates the capabilities that the scripting language must support. Thus, the scripting language has to be tailored to the need. It might be possible to incorporate new parameter mappings on the fly. It might also make it easier to code parameters mappings. However, the grammar and scripting language has to be first defined to derive the benefits out of this approach.

Table 3 Generic translation requirements for the IWF.

Sl. No.	Requirement	Description
1	Translation from/to an AVP to/from an IE	This is a straightforward translation from/to an AVP to/from an IE, which may involve one of the following types: 1. Enumerated to Enumerated 2. Bit string to bit string 3. TBCD string to TBCD string Examples: 1. RAT-Type AVP to usedRAT-Type IE (Enumerated to Enumerated) 2. Supported-Features. Feature-List AVP to supportedFeatures IE (Bit string to bit string) (Feature-List is Unsigned32) 3. MME-Number-for-MT-SMS AVP to mmeNumberforMTSMS IE (TBCD string to TBCD string)
2	Extraction of a part of an AVP to populate an IE	In this, a part of an AVP is extracted and the resulting value is put into an IE. Typically, such an extraction can be performed using string manipulation. Example: User-Name AVP to imsi IE
3	Generation of a parameter using another AVP	In some cases, an IE cannot be directly populated from any AVP. But, possible values for a certain AVP can be used to define a static mapping, locally. At runtime, a lookup is performed for the received AVP value, the mapped value obtained from the configuration, and the IE populated using the mapped value. Example: sgsn-Address IE obtained through the Origin-Host AVP
4	Setting an IE or AVP to present (IE), or constant value	In some cases, for an IE or AVP, there is no corresponding AVP or IE. In such a case, the IE or AVP is populated with a constant value (may be defined through configuration). Further, in case of MAP, an IE can just be encoded as present in the outgoing MAP message. Example: gprsEnhancementsSupportIndicator IE is always encoded to be present – in this case, there is no corresponding incoming AVP.

Table 3 (Continued)

5	Absence of an IE or AVP	In some cases, an IE or AVP is absent, i.e., is not encoded in the message. Example: extensionContainer parameter.
6	Translation of an enumerated type to bit string	In some cases, an enumerated type is used to populate a bit string. Example: RAT-Type AVP to supportedRAT-TypesIndicator IE.
7	Custom function to combine two or more AVPs or IEs into a single IE or AVP	In some cases, two or more AVPs or IEs need to be combined into a single IE or AVP. In such cases, a custom function is used. Example: Terminal-Information.IMEI (UTF8String → 14-digit) and Terminal-Information.Software-Version (UTF8String → 2-digit) AVP values is combined and encoded into add-info.imeisv IE as an OCTET STRING (ASN.1).
8	Encoding presence of an IE based on a certain bit set in an AVP	In some cases, an IE is present if a certain bit is set in an AVP. Example: servingNodeTypeIndicator IE is present if the ULR-Flags AVP has the S6a/S6d Indicator bit set
9	Encoding presence of an IE based on a certain bit NOT set in an AVP	In some cases, an IE is present if a certain bit is NOT set in an AVP. Example: gprsSubscriptionDataNotNeeded IE is present if the ULR-Flags AVP has the GPRS Subscription Data Indicator bit NOT set
10	Encoding presence of an IE based on a binary Enumerated AVP	In some cases, an IE is present based on the enumerated value of an AVP. The enumerated value can be either 0 or 1. That is, if the value is 0, the IE is not present; if the value is 1, the IE is present. Default behavior is the absence of the IE (if the AVP is NOT present). Example: SMS-Only AVP to sms-Only IE

Table 3 (Continued)

11	Translation of an IE based on a condition applied on the incoming message	A condition checks the presence or the absence of an IE and based on the condition a constant value is set to the AVP. Example: In the translation of UpdateGPRSLocationRes/Err to ULA, the Result-Code AVP is mapped to an error code based on the presence or absence of a particular IE.
12	Possible cases: • All the translations are specified at the transaction level. • There are events within a transaction to allow error components and intermediate message components. • There are IE or AVP mappings/settings within events • The translation has a provision to allow state variables, which are valid for the entire transaction state, to store intermediate results in AVPs, IEs or other basic types • The translation allows the setting of state variables to IEs or AVPs within an event • The translation allows set operations from AVPs or IEs to a state variable • The translation allows IE mappings/settings within a transaction or an event	In the case of TCAP based applications, there can be multiple messages exchanged within a transaction. But all might map to a single answer on the Diameter side. Example: In the translation of UpdateG-PRSLocationRes/Err to ULA, if the InsertSubscriberDataArg was sent as part of the same transaction and if the Skip-Subscriber-Data is not present in the previously sent ULR, then the Subscription-Data AVP is populated from the InsertSubscriber-DataArg. But the ULA is sent only when the UpdateGPRSLocationRes/Err is received.

Table 3 (Continued)

13	Translation allows the mapping of an IE of type SEQUENCE to a Grouped AVP and vice-versa	In the ASN.1 notation, an IE that repeats multiple times is denoted as a SEQUENCE. Diameter has the notion of Grouped AVP. The translation allows such a mapping possible. Example: In the translation of InsertSubcriberDataArg to Subscription-Data AVP, Regional-Subscription- Zone-Code is of type OctetString (AVP type) but has a maximum multiplicity of 10. InsertSubcriberDataArg .subscriberData.regionalsubscriptionData, which is a Sequence of Octet String (ASN.1) maps to this.
14	Translation allows the following conversions possible • DNS encoded to UTF8String (displayable string) and vice versa • OCTET STRING (ASN.1) to their Hexadecimal display string equivalent and vice versa • TBCD encoded string to UTF8String (displayable string) and vice versa	DNS encoding of APN, which is nothing but an FQDN, is different from the string representation of a domain name. Diameter uses the plain string representation of the domain name, but this is not the case with legacy protocols. Hence, this is supported. Similar conversions are required for other parameters. Examples: • APN-OI-Replacement from apn-oi-replacement in InsertSubscriberDataArg • 3GPP-Charging-Characteristics is a displayable string (4 bytes), while the charging Characteristics of InsertSubsDataArg is an OCTET STRING (ASN.1) of 2 bytes • IMSI is encoded as UTF8String in Diameter while that is not the case with legacy protocols
15	Translation allows mapping from a SEQUENCE IE to a Grouped AVP and vice versa	Nesting of IEs is possible in ASN.1 with the SEQUENCE type. Similarly, nesting of AVPs is possible in Diameter with the Grouped AVP type. There is a construct that maps the nested IEs to AVPs within a Grouped AVP. This is allowed at multiple levels of nesting. Example: In the translation of InsertSubscriberDataArg to ULA.Subscription-Data, Subscription-Data.AMBR Grouped AVP (with two sub-AVPs) is mapped to ISDArg.eps-SubscriptionData.ambr which is a SEQUENCE type with two parameters

Table 3 (Continued)

16	Translation allows the mapping of a SEQUENCE(nesting) of SEQUENCE (list) to repeated Grouped AVP and vice versa	In the ASN.1 notation, a SEQUENCE can denote grouping of IEs and also denote a list/array. Similarly, in Diameter, Grouped AVPs can be repeated multiple times. It is possible to map a SEQUENCE to a list of Grouped AVPs. This shall be allowed at multiple levels of nesting. Example: In the translation of InsertSubscriberDataArg to ULA.Subscription-Data, Subscription-Data.CSG-Subscription-Data, which is of type repeated Grouped AVP, maps to ISDArg.csg-SubscriptionDataList which is of type list of SEQUENCE
17	Translation allows the following conversions • IE presence to single valued Enumerated AVP • Boolean IE to binary valued Enumerated AVP • IE presence to binary valued Enumerated AVP	Single valued Enumerated AVP has only value (most likely value 0). Binary Enumerated AVP has two values (0 and 1). In ASN.1 an IE can be of type NULL or of type Boolean. All these are used as flags (to enable or disable something). Examples: • In the translation of ISDArg to Subscription-Data, Subscription-Data.Roaming-Restricted-Due-To-Unsupported-Feature with value 0 is set if ISDArg.roamingRestrictedInSgsnDueToUnsupported-Feature IE is present. • Subscription-Data.MDT-User-Consent with values 0 and 1 maps to ISDArg.mdtUserConsent, which is of type Boolean • Subscription-Data.PS-and-SMS-only-Service-Provision is set to1 if the parameter ISDArg. PS-and-SMS-only-Service-Provision is present, else 0 is set.

2.6 Mapping of Routing Parameters

If an interworking is needed after protocol translation, the routing parameters in the incoming request must be mapped to corresponding routing parameters in the outgoing request.

Diameter and TCAP follow different request routing principles. In the case of Diameter, request routing uses the Destination-Realm AVP, Application ID header field and optionally Destination-Host AVP. In the case of TCAP, request routing uses the Signalling Connection Control Part (SCCP) routing principles. This can be either Point Code (PC)-Subsystem Number (SSN) or Global Title (GT) based routing (refer to [4] for CCITT or [5] for ANSI for the corresponding routing principles). Further, a MAP/CAP operation included in a TCAP component has an Application Context Name (ACN).

Here, Diameter S6a/S6d to MAP Gr will be considered in illustrating a typical request routing scenario during interworking.

In a Diameter S6a/S6d request, the User-Name AVP is a mandatory parameter. The User-Name AVP is comprised of three parts and constitutes the subscriber IMSI:

- 3-digit Mobile Country Code (MCC)
- 2 or 3-digit Mobile National Code (MNC)
- Rest, Mobile Subscriber Identification Number (MSIN)

The combination of the MCC and MNC constitute the Home Realm/Domain or the Destination-Realm of the user.

Optionally, the Destination-Host AVP can be used to identify a specific destination. The Destination-Host AVP may be constructed from a range of digits in the MSIN.

Thus, for a request originating on Diameter S6a/S6d, the PC and/or GT is derived from the User-Name AVP. The mapping of the MCC and MNC to PC and/or GT of the HLR is specified in an IWF routing configuration file, along with other configurable SS7 routing parameters such as family (ANSI or CCITT) and National/International indicator.

The Application ID in the Diameter S6a/S6d request header is mapped onto the destination SSN. The destination SSN value corresponds to GSM MAP HLR (6) in the case. Also, the IWF acts on behalf of the SGSN as seen from the HLR. Thus, the origination SSN value corresponds to GSM MAP SGSN (149) in this case – this is also specified in the IWF configuration file.

The outgoing MAP/CAP operation determined from the Diameter Command-Code [2] is used to compute the ACN in the outgoing TCAP dialogue.

The Transaction Manager stores the latest origination context for an IMSI. The origination context refers to the MME/SGSN serving the IMSI currently. This is important for a subsequent server-initiated message such as a Cancel Location initiated from an HLR. This is because for a server-initiated message, the destination must be "exactly" identified, which is obtained from the stored origination context.

In case the User-Name AVP is not present, which may be true for Diameter Gy/S13/S13', the Destination-Realm is directly used to map to the corresponding PC and/or GT – for Diameter Gy to CAP Ge, the PC and/or GT resolves to a Prepaid SCP; for Diameter S13/S13' to MAP Gf, the PC and/or GT resolves to an EIR.

In case of Diameter Gy/S13/S13', if the Destination-Host is not present, the destination node may be identified through a configured Load Balancing scheme.

Mapping of MAP Gr to Diameter S6a/S6d can be easily determined now.

3 Conclusion

Some aspects of Diameter-CAP interworking has not been discussed in this paper. In the case of Diameter Gy, a single Diameter session can span multiple Diameter transactions. In contrast, in the case of Diameter S6a/S6d, a single Diameter session spans one Diameter transaction.

The Diameter–RADIUS interworking has not been discussed at all in this paper. These two interworking scenarios shall be dealt with in detail in a future version of this paper.[1]

References

[1] 3GPP TS 29.305: InterWorking Function (IWF) between MAP based and Diameter based interfaces.
[2] IETF RFC 6733: Diameter Base Protocol.
[3] ITU-T Recommendation Q.771: Functional description of transaction capabilities.
[4] ITU-T Recommendation Q.713: Signalling connection control part formats and codes.
[5] ANSI T1.112: Signalling System Number 7 (SS7) ? Signalling Connection Control Part (SCCP).
[6] 3GPP TS 29.002: Mobile Application Part (MAP) specification.
[7] http://www.3gpp.org/ftp/Specs/archive/29_series/29.002/ASN.1/, Cross-reference listing and fully expanded ASN.1 sources of the MAP protocol.
[8] 3GPP TS 29.272: Mobility Management Entity (MME) and Serving GPRS Support Node (SGSN) related interfaces based on Diameter Protocol.

---

[1] The authors of this paper may however be contacted if information on these interworking scenarios is desired.

Biographies

Arnab Dey is a Technical Lead at Diametriq, LLC. His interests include architecture and design of software, programming, authoring technical white papers in emerging telecom and web technologies, speaking in various forums and mentoring. Dey received a Bachelors in Electrical Engineering from Jadavpur University, India and an Executive Post Graduate Diploma in Business Management from Symbiosis International University, India. Contact him at adey@diametriq.com.

Balaji Rajappa is a Development Engineering Manager at Diametriq, LLC. His interests include network convergence, cloud computing on telecom networks and rich communication suite. Rajappa received a Bachelors in Engineering from Bharathiyar University, India. Contact him at brajappa@diametriq.com.

Lakshman Bana is a Senior Software Engineer. He has worked on telecommunications signaling protocols for more than 12 years. He is currently focusing on mobile applications development and rich communication services. Bana received an MS in Computer Information Systems from Florida Tech, USA. Contact him at lbana@diametriq.com.

Identity Authentication and Capability Based Access Control (IACAC) for the Internet of Things

Parikshit N. Mahalle, Bayu Anggorojati, Neeli R. Prasad
and Ramjee Prasad

Center for TeleInFrastruktur, Aalborg University, Aalborg, Denmark;
e-mail: {pnm, ba, np, prasad}@es.aau.dk

Received 15 September 2012; Accepted 17 February 2013

Abstract

In the last few years the Internet of Things (IoT) has seen widespread application and can be found in each field. Authentication and access control are important and critical functionalities in the context of IoT to enable secure communication between devices. Mobility, dynamic network topology and weak physical security of low power devices in IoT networks are possible sources for security vulnerabilities. It is promising to make an authentication and access control attack resistant and lightweight in a resource constrained and distributed IoT environment. This paper presents the Identity Authentication and Capability based Access Control (IACAC) model with protocol evaluation and performance analysis. To protect IoT from man-in-the-middle, replay and denial of service (Dos) attacks, the concept of capability for access control is introduced. The novelty of this model is that, it presents an integrated approach of authentication and access control for IoT devices. The results of other related study have also been analyzed to validate and support our findings. Finally, the proposed protocol is evaluated by using security protocol verification tool and verification results shows that IACAC is secure against aforementioned attacks. This paper also discusses performance analysis of the protocol in terms of computational time compared to other

Journal of Cyber Security and Mobility, Vol. 1, 309–348.

existing solutions. Furthermore, this paper addresses challenges in IoT and security attacks are modelled with the use cases to give an actual view of IoT networks.

Keywords: access control, authentication, capability, Internet of Things.

1 Introduction

In the Internet of Things (IoT) [1, 2], every virtual and physical entity is communicable, addressable and is accessible through the Internet. These virtual and physical entities produce seamless communication and seamless service collaborating with users and other devices creating service oriented networks. The IoT is an emerging paradigm and makes the world of computing fully ubiquitous creating UbiComp, a term initially coined by Mark Weiser [3]. Due to rapid development in Radio Frequency Identification (RFID) [4] technology, Wireless Sensor Networks (WSN), actuators and mobile communication, it is possible to realize the IoT due to ubiquitous interactions between things and devices in an "anytime, anywhere and anything" form.

Any "thing" with sensing, communication and computation capability helps us to realize the IoT vision and there are many application areas possible due to these smart thing or objects. These IoT applications are categorized in four domains in [5]:

- Personal and Home – includes individual homes [6].
- Enterprise – includes scales of community [7].
- Utilities – includes national and regional scales [8].
- Mobile – includes IoT applications spread across multi-domain due to distributed connectivity and scale [9].

An example application area is intelligent home environment (personal) which mainly consists of places full of things that will interact with each other at different levels. There are different kinds of sensors and devices that use heterogeneous technologies; low bandwidth meshes networking based (such as ZigBee and Z-Wave) or other high bandwidth demanding (such as Bluetooth, WiFi, 4G or UWB) providing 24×7 monitoring or entertainment services. Other application area includes nomadic access to services where accessible services are discovered according to the user's identity and profile with the help of a mobile device. eHealth is the most important application of IoT, where sensors, actuators, RFID tags, etc., are applied in the health sector to facilitate ease of life service across geographic and time barriers.

The main challenges in these application areas are to ensure that ubiquitous access to services and monitoring data is granted to identities that fulfil the access control rules for identity management, heterogeneous device interaction, authorization, mutual authentication and secure delegation from a mobile device, and the secure data access. Securing user interactions with IoT is essential if the notion of "things everywhere" is to succeed. In such a scenario, security and privacy are two key challenges [10] that will determine the success or failure of a connected world.

The remainder of this paper is organized as follows: Section 2 presents the technological challenges and security challenges that need to be addressed to realize the notion of IoT. Section 3 presents the related works in authentication and access control. Threat analysis and attack modelling is presented in Section 4. Section 5 presents the proposed scheme for mutual authentication and access control. Evaluation of the proposed scheme using protocol verification tool and performance analysis is presented in Section 6. Finally, Section 7 concludes the paper with future work.

2 Challenges

As outlined in the scenarios and the applications above, it is clear that we are transforming from an Internet of computers to the Internet of things with device to device communication. In order to make the IoT services available at low cost with a large number of devices communicating to each other, there are many challenges to overcome. These challenges are divided into two categories in this paper as:

- *Technological challenges* – These challenges are related to underlined wireless technologies, energy, scalability, distributed and dynamic nature of IoT and ubiquitous interactions.
- *Security challenges* – These challenges are related to security services like authentication, privacy, trustworthiness and confidentiality. Security challenges also include heterogeneous communication and end-to-end security.

2.1 Technological Challenges

- *Wireless Communication*: IoT significantly uses convergence of established wireless technologies such as GSM, UMTS, Wi-Fi, Bluetooth and WPAN. These underlined wireless technologies use different standards and have different communication bandwidth requirement. This convergence also creates serious interoperability issues.

- *Scalability*: Unbounded number of devices creates the larger scope and scalability in IoT than conventional communication networks. IoT covers large application areas like a home environment where number of devices are relatively small in number to a factory or building that has a large number of devices offering multiple services to the users. IPV6 is one attempt to accommodate as many numbers of devices and things in IoT.

- *Energy*: IoT consist of constrained objects which do not have enough power, memory and computation capabilities. Designing lightweight protocols for IoT which minimize energy consumption is very important as compared to conventional protocols running on devices with sufficient resources.

- *Distributed and Dynamic Nature*: In IoT, things can interact with other things at any time, from anywhere and in any way independent of the location. As the IoT networks are distributed in nature, designing protocols for them is a challenging task. The objects interact dynamically and hence appropriate services for the objects must be automatically identified. In addition to this, the mobility/roaming of the objects is another important challenge.

- *Identification*: In the IoT, things include variety of objects like computers, sensor nodes, people, vehicles, medicines, books, etc. These things should be uniquely identified for the addressing capabilities and for providing a means to communicate with each other. After verifying the identities of things, we call these uniquely identified things as objects. Different identity schemes have been proposed for the IoT and it is predicted that it is dubious to have common identification schemes globally. Identification schemes like RFID Object Identifier, EPCglobal, Short-OID and Near Field Communications Forum, IPV4, IPV6 and E.164 have been studied in the literature. These addressing methods/principles are highly depends on the underlined access technology, thus it is challenging to have many different addressing protocols for varied underline access technologies.

2.2 Security Challenges

- *Privacy*: Privacy is one of the most sensitive areas in the context of IoT. In IoT, all objects are connected to the Internet and they communicate with each other over the Internet. Hence the privacy issue is critical. As the Internet gets diversified with new types of devices and heterogeneous networks, IoT users and devices have to access the digital world with wide range of methods and protocols. Further, as ownership of these devices by the users does not exist, the issue of privacy is aggravated.

- *Identity Management*: Due to the scale of economics in the IoT, un-bounded numbers of things or objects are involved in accessing IoT networks and communicating with each other. Hence, efficient and light-weight identity management schemes are required. In addition to this, the distributed nature of IoT makes this problem more challenging.

- *Trust*: Trust is an essential and integral factor to consider when implementing IoT. In an uncertain IoT environment, trust plays an important role in establishing secure communication between things. There should be an effective mechanism to define trust in a dynamic and collaborative IoT environment. It is also important to provide context aware trust management for varied IoT applications.

- *End-to-End Security*: End-to-end security measures between IoT devices and Internet hosts are equally important. Applying cryptographic schemes for encryption and authentication codes to a packet is not sufficient for the resource constrained IoT. Hence future research is required into efficient end-to-end security measures between IoT and the Internet.

- *Authentication and Access Control*: Authentication is identity establishment between communicating parties. Authentication and access control is important to establish secure communication between multiple devices and services. Interoperability and backward compatibility are the two key issues to be addressed. For example, in Wi-Fi roaming, devices use UMTS at the core networks.

- *Attack Resistant Security Solution*: Due to diversity of devices and end users, there should be attack resistant and lightweight security solutions. All the devices in IoT have low memory and limited computation resources, thus they are vulnerable to resource enervation attack. When the devices join and commissioned into the network, keying material, security and domain parameters could be eavesdropped. Possible external attacks like denial of service attack, flood attack, etc., on device and mitigation plan to address these attacks is another big challenge.

3 Related Works

There is ongoing research in the field of authentication and access control. This section presents state of the art in authentication and access control.

3.1 Authentication

There is much research done in the area of securing IoT. There is closely related work done in the MAGNET project [11, 12] where security associations take place with increased communication overhead and authentication is left unaddressed. The authors presented a distributed access control solution based on security profiles but attack resistance is not explored. In [13, 14], the authors have presented an ECC based authentication protocol but the major disadvantage is that it is not Denial of Service (DoS) attack resistant. As there are billions of devices in IoT, resistance to DoS attack is of vital importance. In [15], the author addressed the problem of secure communication and authentication based on a shared key and is applicable to limited location and cannot be used for wide area. It addresses peer to peer authentication but cannot be extended to a resource constrained environment.

There has been lot of debate about which of the cryptographic primitives like public key or private key is suitable for the IoT. Most of the research has mainly focused in areas like WSN and applications like healthcare and smart home. Many security mechanisms have been proposed based on private key cryptographic primitives due to fast computation and energy efficiency. Scalability problem and memory requirement to store keys makes it inefficient for heterogeneous devices in IoT.

A public key cryptography based solution overcomes these challenges because of its high scalability, low memory requirements and no requirement of key pre-distribution infrastructure. In [16], the author presented ECC based mutual authentication protocol for IoT using hash functions. Mutual authentication is achieved between terminal node and platform using secret key cryptosystem introducing the problem of key management and storage. Self-certified keys cryptosystem based distributed user authentication scheme for WSN is presented in [17], where only user nodes are authenticated. However, this is not lightweight solution for IoT. In [18], the author presented an authentication with parameter passing during the handshake. The handshake process is time consuming and based on symmetric key cryptography with more memory requirement for large prime numbers. Efficient identification and authentication is presented in [19] and is based on the signal properties of the node but it is not suitable for mobile nodes. The direction of the signal is considered as a parameter for node authentication but it takes more time to decide the signal direction with more memory and computations involved. In [20], cluster based authentication is proposed which is most suited for the futuristic IoT, but an attacker can get hold of the distribution of system key pairs and cluster key. Generation of random numbers and signatures creates considerable computational overhead consuming memory resources.

Mobility is very important aspect of mobile and wireless communication and essentially in the context of IoT. With the heterogeneous network topologies like Wi-Fi, LTE and WiMax, authenticated service delivery with proper access control in place on the fly is a big challenge. Wireless Internet Service Provider roaming (WISPr) [21, 22] is an architecture, which proposes detailed specifications for allowing inter-operator roaming for Wi-Fi clients. Roaming functionalities in the vendor devices is based on the IANA Private Enterprise Number (PEN). WISPr enables users for roaming between different wireless Internet service providers. WISPr uses Remote Authentication Dial in User Service (RADIUS) [23] to provide centralized authentication and authorization. Analysis and security vulnerabilities of RADIUS have been discussed in [24] due to its centralized nature. Extensible Authentication Protocol (EAP) [25] is authentication framework being used in Wi-Fi. Security assessments of EAP have been discussed in [26] and explored many weakness points. Especially EAP do not address mutual authentication and not resistant to replay attack [26]. Key replication and replay attack on Authenticated Key Agreement (AKA) have been presented in [27] which clearly shows that there is even an identity is associated with AKA, it is prone to attack. Comparative studies on authentication and key agreement methods for 802.11 wireless LANs is presented in [28]. Weaknesses and security assessment of various authentication methods in the context of wireless networks is very well presented in [28]. General requirements for authentication and key agreement are classified into three mutually exclusive sets as: mandatory, recommended and additional requirements. A multi-layer agreement protocol is also proposed in [28]. This state of the art in mobile and Wi-Fi environment clearly shows that there is a need of flexible and secure authentication scheme.

State of the art evaluation is shown in Table 1. Related work is summarized based on the parameters like mutual authentication, lightweight solution, resistant to attacks, distributed nature and access control solution. Recent related work in the area of authentication for IoT is considered for the evaluation and is presented below.

Table 1 State of the art evaluation summary.

Solutions	Mutual Authentication	Lightweight Solution	Attack Resistant			Distributed Nature	Access Control
			Dos	Man in middle	Replay		
Ubiquitous Access Control in MAGNET [11, 12]	No	No	No	No	No	Yes	Yes
ECC based Authentication in RFID [13, 14]	Yes	Yes	No	No	No	Yes	No
Authentication in Ad-hoc Wireless Networks [15]	No	No	Yes	Yes	Yes	No	No
Authentication in IoT [16]	Yes	Yes	No	Yes	Yes	Yes	No
Authentication in WSN [17]	No	No	No	No	No	Yes	No
Progressive Authentication in Ad-hoc Networks [18]	Yes	Yes	No	Yes	Yes	No	No
Peer Identification and Authentication [19]	Yes	No	No	No	No	Yes	No
Authentication in Ad-hoc Networks [20]	Yes	No	No	No	No	Yes	No

From Table 1, it is clear that not all existing solutions for authentication fulfil each and every requirement for IoT. The NO block in Table 1 represents the respective feature unavailability in the corresponding solution. Evaluation summary of the state of the art shows that all existing authentication solution in Wi-Fi environment and in the context of IoT do not address all the requirements like attack resistant, mobility and lightweight solution and mutual authentication.

3.2 Access Control

Controlling access to information or resources is usually done by defining access control rules, which decide who is allowed to access what and who is not. These rules take different forms such as RBACs, ACLs, policies, and so on. Before the development of standards based policy languages, interoperability was a major concern. It was with the emergence of the XACML proposal

[29], defined by OASIS, that identity management developers started thinking about how to make use of such standards based languages to define the set of policies, and to provide more standard solutions. In the IoT world, such standards based solutions are imperative due the distributed nature of the problem. XACML includes an XACML delegation profile in order to support administrative and dynamic delegation. The purpose of this profile is to specify how to express permissions about the right to issue policies and to verify issued policies against these permissions. This profile led to an identity federation scenario, is the key element upon the management of delegation policies. At the moment there is not a solution to define the relationship among the involved institution in a service interaction, neither a way to combine the decision taken by different organizations. There is currently no standard proposal related with the establishment of agreement at organization, federation or other trust domains levels. Examples of this kind of policies could be common information representation format, security requirements, levels of trusts, etc. This policy can be taken as a starting point for the definition of a negotiation mechanism about capabilities and policies, independently of the kind of entity involved on it. Policy and Charging Control (PCC) in LTE enables centralized mechanism for charging control and service-aware quality of service. PCC operates in S9 interface and consist of Policy and Charging Rule Function (PCRF) which controls the policies dynamically based on subscriptions and sessions between home PCRF and visited PCRF. Consider the scenarios of heterogeneous home M2M network in IoT based on LTE/4G. In this scenario, home gateway proactively and adaptively interacts with the surrounding radios in order to connect to home network and in turn to the external networks. Security policies protect the home M2M network from possible external attack via trusted access control and networked encryption technique.

Although XACML was the starting point towards the definition of standard policies, it is only focused on the resource access control type of policy. More or less at the same time, other kind of policies emerged to cover specific aspects for identity management, for example P3P [30], to define online privacy release information policies between end users and services. Current systems have incorporated these kinds of standard policies in some way, for example Shibboleth [31] and Liberty Alliance [32] providing definition of access control policies by means of XACML. However, there is a need to define policies in a standard way in the next generation of policy-driven systems when distributed scenarios in the IoT domain are considered.

It is equally important to discuss the state of the art in access control solutions. Traditionally, access control is represented by Access Control Matrix (ACM), in which the column of ACM is basically a list of objects or resources to be accessed and the row is a list of subjects or whoever wants to access the resource. From this ACM, two traditional access control models exist, i.e. Access Control List (ACL) and capability based access control. Many scientists [33, 34] have made comparisons between ACL and capability based access control and the conclusion is that ACL suffers from a confused deputy problem and other security threats while it is not the case in the capability based access control. Moreover, ACL is not scalable being centralized in nature and also it is prone to single point of failure. It cannot support different level of granularity and revocation is time consuming with lack of security. However, several drawbacks have been identified in applying the original concept of capability based model into access control model as it is to IoT. Gong [35] pointed out two major drawbacks of classical capability based model namely the capability propagation and revocation, and provide solutions to them by proposing a so called Secure Identity based Capability System (ICAP). Yet, Gong [35] did not clearly describe the security policy that is used in the capability creation and propagation. It also did not consider context information in making access control decision upon access request from a subject or user.

Nowadays Internet and web based applications are widely used and different types of access control models have appeared, such as Role Based Access Control (RBAC), Context Aware Access Control (CWAC), Policy Based Access Control, etc. Among others, RBAC is considered to be the most famous access control method in terms of the usage and implementation. In [36–42] extensions of the RBAC model are presented. As mentioned in [34], the RBAC model is essentially a variation of identity based access control to which ACL is sometimes referred, which seeks to address the burdens of client identification. Therefore, the RBAC model is still vulnerable to confused deputy problem as is the case with an ACL based model. Moreover, due to the role based structure in RBAC, it is not a generic model. As access permissions to the entities can be assigned through roles only, it has limited granularity. Scalability and delegation is critical in RBAC and it is not time efficient for micro level access. In [37], the authors presented General Temporal RBAC (GTRBAC), a RBAC based model that capable in expressing a wide range of temporal constraints, in particular periodic as well as duration constraints on roles, user-role assignments, and role-permission assignments. An example of GTRBAC's usage in the real world application is in defining access rights to employees in a company who work based on shifts, e.g. morning, afternoon, and night shift, and also for people who work on short term contracts, and many others. However, it is not able to describe the limitation of any context other than periodic or time duration. Bhatti et al. [38] addressed the issues in XACML as well as GTRBAC with emphasize in formal definition of context, and introduction of trust model with RBAC and XML main features. However, the scope is only limited to web service environments and hence not really suitable to the IoT. Privacy aware RBAC is presented in [39] and compared with XACML but its application to IoT is unclear.

In [40–42], the authors addressed the issue of role and/or permission delegation based on the RBAC model. However, unlike Barka and Sandhu [40, 41], Hasebe et al. [42] considered delegation of roles and permissions in a cross-domain environment by using capability, and thus it is called Capability RBAC (CRBAC) model. The main idea of CRBAC is essentially similar to what has been proposed in [35], i.e. by using capability transfer or propagation in order to delegate roles or permissions. However, the main aim of using capability is limited to delegation only, thus it does not exploit the capability fully. Moreover, explanation of the revocation of delegation or capability transfer was not discussed, plus other drawbacks related to [39] and RBAC as described earlier are also applicable here.

In CWAC [43], the surrounding context of the subject and/or object is considered to provide access. Scalability is again a problem with CWAC. Delegation and revocation is not supported completely in CWAC. In CRBAC [44], context is integrated with RBAC dynamically. Context is defined as characterization of surrounding entities for performing appropriate actions. Improper association of context and role results in scalability and time inefficiency. Further, the delegation is not simple due to context dependency. There are many examples like context aware patient information system and context aware music player where applying role based access control is a cumbersome process.

Comparison of these access control models is shown in Table 2. Comparison is based on functional parameters such as generic nature, scalability, granularity, delegation, time efficiency, and security.

State of the art for authentication and access control shows that there is no integrated protocol for authentication and access control. The objective is to achieve mutual identity establishment, i.e. authentication and once authenticated, access control will take place. This paper proposes a new method of authentication of devices and access control for the IoT resources using public key approach with scalability and less memory requirements. The most

Table 2 Comparison of different access control models.

Models	Generic	Scalable	Granular	Delegation	Time Efficient	Security
ACL	Yes	No	No	No	No	No
RBAC	No	No	Yes	Yes	No	No
CWAC	Yes	No	Yes	No	No	No
CRBAC	Yes	No	Yes	Yes	No	No
CCAAC	Yes	Yes	Yes	Yes	Yes	Yes

important design issue of IoT is the mobility of heterogeneous devices and proposed scheme works efficiently for this need.

4 Threats and Attacks Modelling

An important endeavour of this paper is to model the activities of IoT attacks to understand the sequence of actions taking place when the attacks are happening. The modelling of the security attacks helps to understand an actual view of the IoT networks and enable us to decide the mitigation plans.

In the IoT, the possible communications are device to device, human to device and human to human giving connection between heterogeneous entities or networks. Figure 1 presents general use case of IoT, where MobileEntity(x): A mobile device represents an entity, i.e. any device in the network which communicates with other entities of same type or of different type via Internet or direct. MobileEntity 1, 2, 3 represent three different and most probable scenarios in the system of communication. Use Cases are self-explanatory and attackers are at the top of the diagram.

- *Man-in-the-Middle Attack*: When the devices are commissioned into a network, keying material,security and domain parameters could be eavesdropped. Keying material can reveal the secret key between devices and authenticity of the communication channel could be compromised. Man-in-the-middle attack is one type of eavesdropping possible in the commissioning phase of devices to IoT. The key establishment protocol is vulnerable to man-in-the-middle attack and can compromise device authentication as devices usually do not have prior knowledge about each other. As device authentication involves exchange of device identities, identity theft is possible due to man-in-the-middle attack. Sample use case for man-in-the-middle attack is shown in Figure 2.

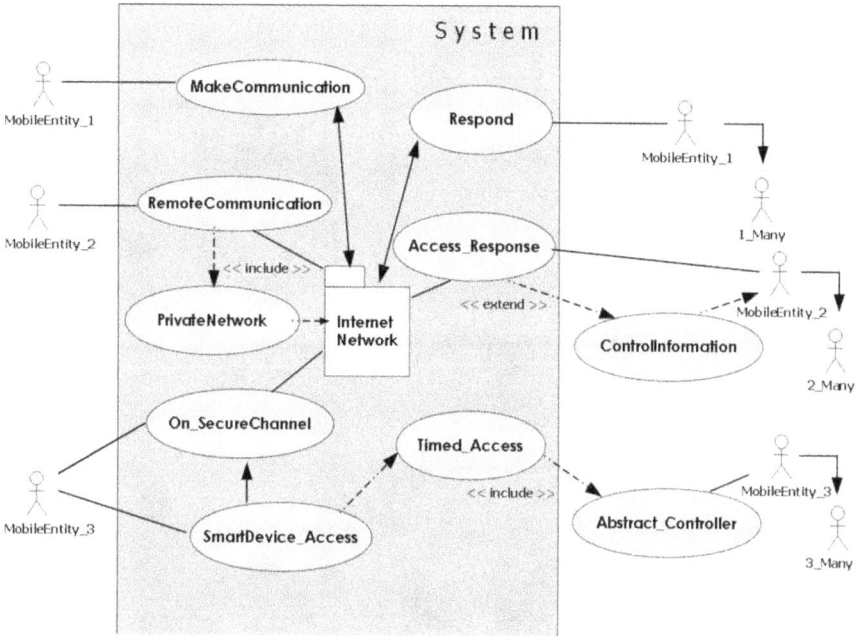

Figure 1 IoT use case.

- *Denial of Service Attack*: All the devices in IoT have low memory and limited computation resources, thus they are vulnerable to resource enervation attack. Attackers can send messages or requests to specific device so as to consume their resources. This attack is more daunting in IoT since attacker might be single in number and resource constrained devices are large in numbers. DoS attack is also possible due to man-in-the-middle attack. Sample use case of DoS in IoT scenario is shown in Figure 2.
- *Replay Attack*: During the exchange of identity related information or other credentials in IoT, this information can be spoofed, altered or re-played to repel network traffic. This causes a very serious replay attack. Replay attack is essentially one form of active man-in-the-middle attack. Our solution prevents replay attacks by maintaining the freshness of random number, for example by using time stamp or nonce by including Message Authentication Code (MAC) as well. A sample use case is shown in Figure 2.

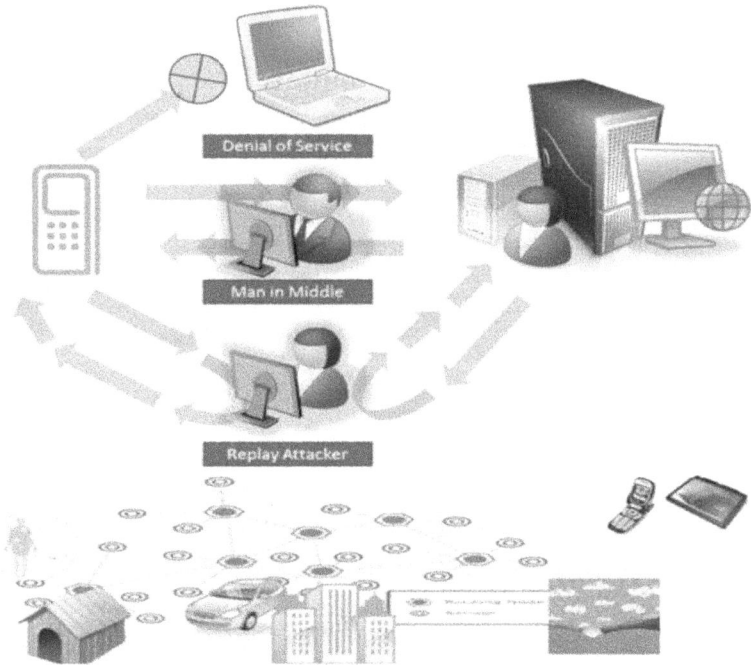

Figure 2   IoT security attacks modelling.

For this purpose, authentication and access control are main security issues which are to be addressed. This paper presents an integrated lightweight solution for authentication and access control with the protocol evaluation.

5 Proposed IACAC Model

As stated earlier, mobility is very important aspect of wireless communication and essentially in the context of IoT. With the heterogeneous network topologies like Wi-Fi, LTE and WiMax, authenticated service delivery with proper access control is major problem to be addressed. Wireless Internet Service Provider roaming (WISPr) [21, 22] and RADIUS [23] are the existing solutions to provide centralized authentication and authorization in Wi-Fi. Related work in security analysis [24–28] shows that there is a need of attack resistant and integrated approach for authentication and access control. Security flaws of authentication and access control protocols have been studied in [45] in the context of mobile communication. Required goals for authentication protocols between mobile entities and fixed networks have been presented in [46], which includes mutual authentication, confidentiality and the attack resistance. Hybrid cryptography based authentication scheme is presented in [47], which is prone to attack on key share and replay attack. Aziz and Diffie [48] proposed mobile authentication and key agreement protocol based on public key cryptography, but it is prone to impersonation attack [49]. The Wong–Chan mobile authentication protocol [50] is vulnerable to DoS attack where malicious initiator can disturb the execution of protocol through bogus request. This makes the Wong–Chan scheme not suitable for resource constrained environment.

This paper presents an Identity Authentication and Capability based Access Control (IACAC) scheme for the IoT to replace the existing schemes. IACAC is compatible with underline access technologies like Bluetooth, 4G, WiMax and Wi-Fi. IACAC presented in this paper is implemented in a Wi-Fi environment and the performance results are discussed in next sections.

The algorithm presented in this paper addresses both authentication and access control which are divided into three parts:

- Secret key generation based on Elliptical Curve Cryptography-Diffie Hellman algorithm (ECCDH),
- Identity establishment,
- Capability creation for access control.

5.1 Secret Key Generation Based on ECCDH and Identity Establishment for Authentication

There is considerable interest in ECC for IoT security [51]. It has advantages of small key size and low computation overhead. It uses public key cryptography approach based on elliptic curve on finite fields. ECCDH [51] is a symmetric key agreement protocol that allows two devices that have no prior knowledge about each other to establish a shared secret key which can be used in any security algorithm. Using this public parameter and own private parameter, these parties can calculate the shared secret. Any third party, who does not have access to the private details of each device, cannot calculate the shared secret from available public information. All devices joining IoT share key pairs during the bootstrapping. The IACAC scheme presented in this paper is also applicable to security bootstrapping. Security bootstrapping is the process by which devices join the IoT with respect to location and time. It includes device authentication along with credential transfer. Protocol uses one or more trusted Key Distribution Center (KDC) to generate domain paramet-

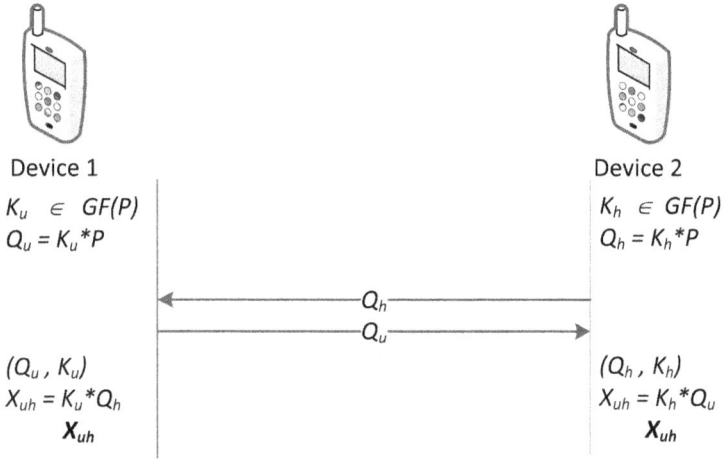

Device 1
$K_u \in GF(P)$
$Q_u = K_u * P$

Q_h
Q_u

(Q_u, K_u)
$X_{uh} = K_u * Q_h$
$\mathbf{X_{uh}}$

Device 2
$K_h \in GF(P)$
$Q_h = K_h * P$

(Q_h, K_h)
$X_{uh} = K_h * Q_u$
$\mathbf{X_{uh}}$

Figure 3 ECCDH for establishing shared secret key.

ers and other security material and important part is this KDC is not required to be online always. Initially KDC randomly selects particular elliptic curve over finite field $GF(p)$ where p is a prime and makes base point P with large order q (where q is also prime). KDC then picks random $x \in GF(p)$ as a private key and publishes corresponding public key $Q = x \times P$. KDC generates random number $Ki \in GF(p)$ as a private key for device i and generates corresponding public key $Q_i = Ki \times P$. The key pair $\{Q_i, K_i\}$ is given to device i. With the increasing number of devices, KDC can generate an ECC key pair based on base point P for any number of devices as it is rich in terms of resources as compared to other devices in IoT. These ECC key pairs will be used to share common secret key for secure communication using ECCDH and is explained below. Steps of aforementioned ECCDH are shown presented in Figure 3.

The assumption here is that ECC is running at trusted KDC. There is an agreement on system based point P and generate (Q_u, K_u) and (Q_h, K_h) pairs where Q_u is the Public key of Device 1; K_u is the secret key of Device 1; Q_h is the public key of Device 2; and K_h is the secret key of Device 2. Furthermore, P is large prime number over $GF(P)$ and generations of above keys are shown in Figure 3.

No parameter is disclosed in this process of establishing a shared secret key other than domain parameter P and public keys. In this paper, we consider sensor nodes as a device, because the functionalities and operational principle of wireless sensor networks makes it an appropriate and mandatory candidate of the IoT.

5.2 Protocol for Identity Authentication

5.2.1 One Way Authentication

One way authentication authenticates Device 1 to Device 2 and is explained below. As per above ECCDH, both Device 1 and Device 2 have X_{uh} as a common secret key. Device 1 selects $r \in GF(P)$ which will be used to create session key. T_u is generated as a time stamp by Device 1. It is assumed that synchronization is taken care using appropriate mechanism. The secret key is created by Device 1 as $L = h(X_{uh} \oplus T_u)$. Then Device 1 encrypts r with secret key L as $R = E_L(r)$ and encrypts T_u by X_{uh} as $T_{us} = EX_{uh}(T_u)$. After this Device 1 builds a Message Authentication Code (MAC) value as $MAC_1 = MAC(X_{uh}, R\|ICAP_1)$ where $ICAP_1$ is a data structure representing an identity based capability for this Device 1 giving access rigts. Details about ICAP are given in the same section below. Now Device 1 sends the following parameters to Device 2 directly or through gateway node/coordination node or access point as (R, T_{us}, MAC_1). Device 2 generates its current time stamp as T current and Device 2 will decrypt T_{us} to get T_u and compare it with $T_{current}$. If $T_{current} > T_u$, it is valid. Now Device 2 calculates L and decrypts R to get r. Device 2 also calculates the MAC_1' and it will verify this with the MAC_1 received from Device 1. If valid, then Device 1 is authentic to Device 2. Device 1 also matches the $ICAP_1$ received with $ICAP_2$ stored at Device 2. If Device 2 gets a match with R, MAC_1, T_{us}, then Device 1 is authenticated to Device 2. This protocol is presented in Figure 4.

5.2.2 Mutual Authentication

This part of authentication authenticates Device 2 to Device 1, and is explained in Figure 5. Device 2 builds a MAC as $MAC_2 = MAC(r\|ICAP_2)$ and also encrypts r with X_{uh} as $R' = EX_{uh}(r)$. Device 2 sends (R', MAC_2) to Device 1. Device 1 verifies MAC_2 and decrypts R' and compares received r with this r (denoted as r' and r'' in Figure 5). If a match is found, Device 2 is also authenticated to Device 1 and communication and access will be granted based on the $ICAP_2$. This protocol achieves both mutual authentication along with capability based access control in secure way.

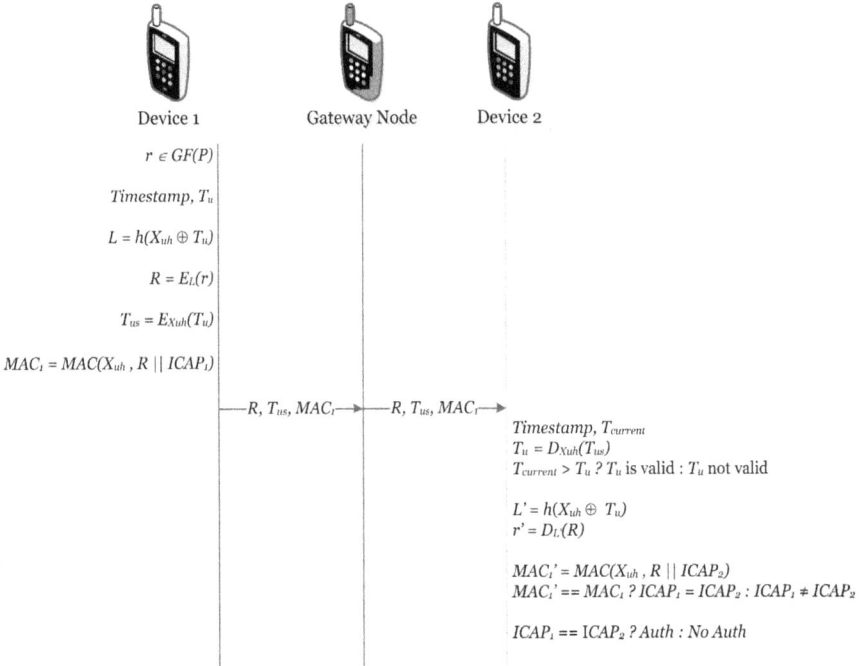

$$r \in GF(P)$$
$$Timestamp, T_u$$
$$L = h(X_{uh} \oplus T_u)$$
$$R = E_L(r)$$
$$T_{us} = E_{Xuh}(T_u)$$
$$MAC_1 = MAC(X_{uh}, R \parallel ICAP_1)$$

$$\xrightarrow{R, T_{us}, MAC_1} \xrightarrow{R, T_{us}, MAC_1}$$

$$Timestamp, T_{current}$$
$$T_u = D_{Xuh}(T_{us})$$
$$T_{current} > T_u ? T_u \text{ is valid} : T_u \text{ not valid}$$

$$L' = h(X_{uh} \oplus T_u)$$
$$r' = D_{L'}(R)$$

$$MAC_1' = MAC(X_{uh}, R \parallel ICAP_2)$$
$$MAC_1' == MAC_1 ? ICAP_1 = ICAP_2 : ICAP_1 \neq ICAP_2$$

$$ICAP_1 == ICAP_2 ? Auth : No\ Auth$$

Figure 4   One way authentication protocol.

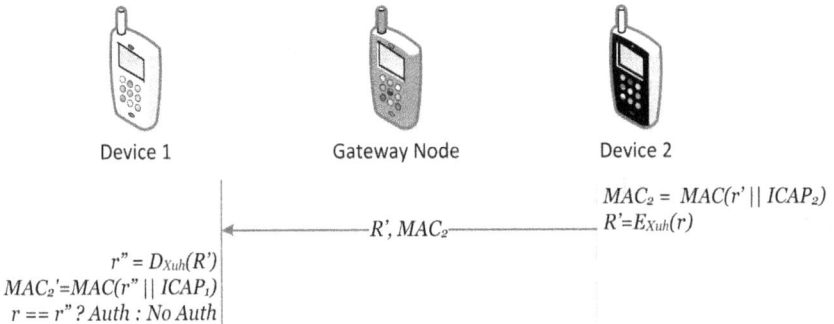

$$MAC_2 = MAC(r' \parallel ICAP_2)$$
$$R' = E_{Xuh}(r)$$

$$\xleftarrow{R', MAC_2}$$

$$r'' = D_{Xuh}(R')$$
$$MAC_2' = MAC(r'' \parallel ICAP_1)$$
$$r == r'' ? Auth : No\ Auth$$

Figure 5   Protocol for mutual authentication.

5.2.3 Capability Creation for Access Control

Conceptually, a capability is a token that gives permission to access device. A capability is implemented as a data structure that contains two items of information: a unique device identifier and access rights. A capability structure is presented in Figure 6. For simplicity, it is sufficient to examine the

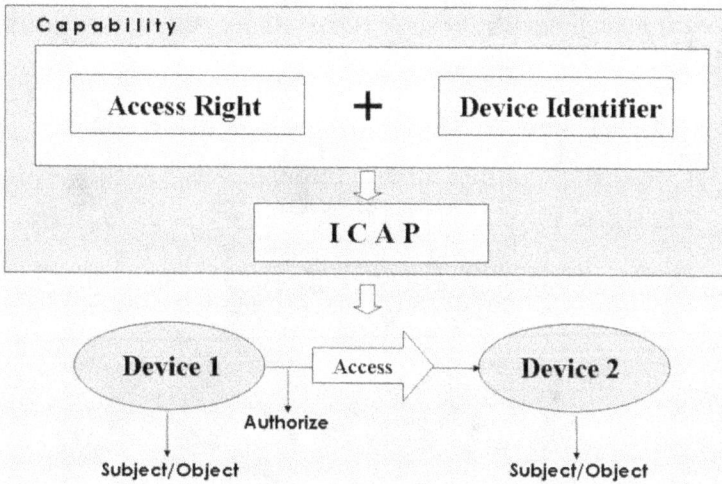

Figure 6 Capability structure.

case where a capability describes a set of access rights for the device. The device which may also contain security attributes such as access rights or other access control information. The ICAP [35] was essentially extending the capability system concept, in which the capability is used by any user or subject that wants to get access to a certain device or resource.

If the capability that is presented by the subject matches with the capability that is stored in the device or an entity that manages the device, access is granted. However, unlike the classical capability based system, ICAP introduced the identity of subject or user in its operation. In this way, it claimed to reduce the number of capabilities stored in the so-called "Object Server", "Gateway" or "Access Point" and thus offers more scalability. Moreover, it has better control in capability propagation which provides more efficient access later on. The ICAP structure and how capability is used for access control is shown in Figure 6. ICAP is represented as

$$ICAP = (ID, AR, Rnd) \tag{1}$$

where ID presents the device identifier; AR the set of access rights for the device with device identifier as ID; and Rnd the random number to prevent forgery and is a result of one way hash function as: $Rnd = f(ID, AR)$. In IACAC, access rights are sent in the form of a MAC value in the authentication process. Implementation works in two stages. First, the devices are

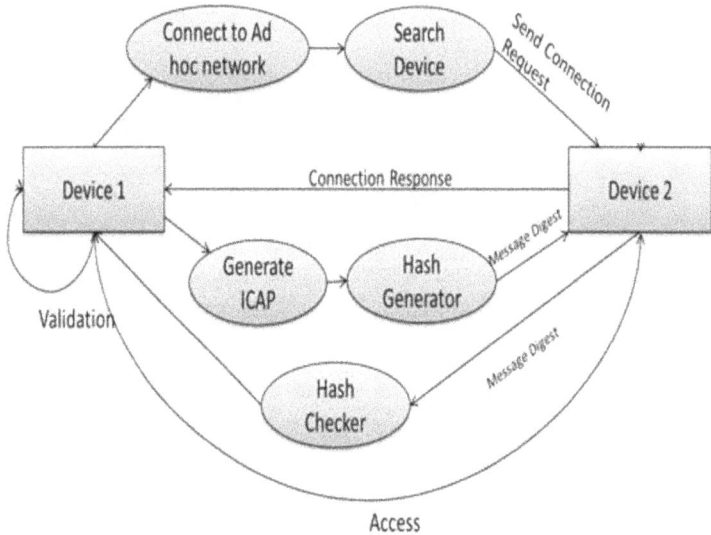

Figure 7 High level functioning of CAC.

connected with each other through the use of an access point and then the capability based access is allowed to the other device through Capability based Access Control (CAC). Each communication that is to be established is verified by its capability access. Only after the capability verification the devices are able to communicate with each other. Any device wants to communicate with other device is able to initiate the communication by sending the request to a specific device. The second stage is to verify whether that requesting device is having the capability to communicate with called device. This access right gets checked using the capability of that device which is associated with every device. For sending capability message digest using SHA-1 is generated for each device as stated earlier and the remote device will check its validity using SHA-1. Figure 7 depicts high level functioning of CAC.

The complete CAC scheme is presented in Figure 8. Figure 8 shows access based on CAC between two Wi-Fi devices. In this paper, we treat all devices as subjects and resources to be accessed as objects. In this implementation of CAC, file is considered as object for access. Access rights (AR) is given as

$$AR \in \{Read, Write, NULL\} \tag{2}$$

AR can either be {Read}, {Write}, {Read, Write} or {NULL}. If AR = {NULL}, the permission to access particular object is not allowed. Once the capability is verified against forgery, both devices are able to perform an operation as specified in capability and access is granted. As any device can perform only those operations as specified in capability, principle of least privilege is supported to a large extent.

6 IACAC Evaluation and Analysis

6.1 Protocol Evaluation

The evaluation will focus on identity authentication in terms of one way and mutual as the most important processes in the authentication. The Automated Validation of Internet Security Protocols and Applications (AVISPA) tool [52] based on the Dolev–Yao model [53] is used for model and protocol verification. We implement the aforementioned protocol in stages. The first stage of protocol authenticates Device 1 to Device 2, i.e. one way authentication, and the second stage is for mutual authentication, i.e. authenticates Device 2 to Device 1. The verification results are described below.

6.1.1 Evaluation Procedure

In order to carry out the evaluation using AVISPA some assumptions are made. Both devices have already obtained ECC based shared key using Diffie–Hellman (ECCDH). As stated earlier, assumption here is that KDC is secure and trusted. Complete protocol evaluation is presented in the following model:

$$D_1 \rightarrow D_2 : [R, T_{us}, \text{MAC}_1]; [\{r\}_L, \{T_u\}\_X_{uh}, \text{RND}_1]$$

$$D_1 \leftarrow D_2 : [R', \text{MAC}_2]; [\{r\}\_X_{uh}, \text{RND}_2]$$

where D_1 is Device 1; D_2 is Device 2; $\{ \}\_$ presents a symbol of encryption; T_u is the timestamp generated as a nonce; X_{uh} is a shared key between D_1 and D_2 using ECCDH; r is some value $x \in GF(p)$; RND_1 is the MAC value of X_{uh}, R and ICAP_1 where ICAP is the result of a one way hash function $f(\text{Device\_ID, Access Rights, Rnd})$, Rnd is a random number generated to prevent forgery; RND_2 is the MAC value of r and ICAP_2; and L presents the result of one way hash function (XOR of X_{uh} and T_u).

Besides this, the Dolev–Yao intruder model has been introduced in the evaluation. The intruder is assumed to have knowledge of the following:

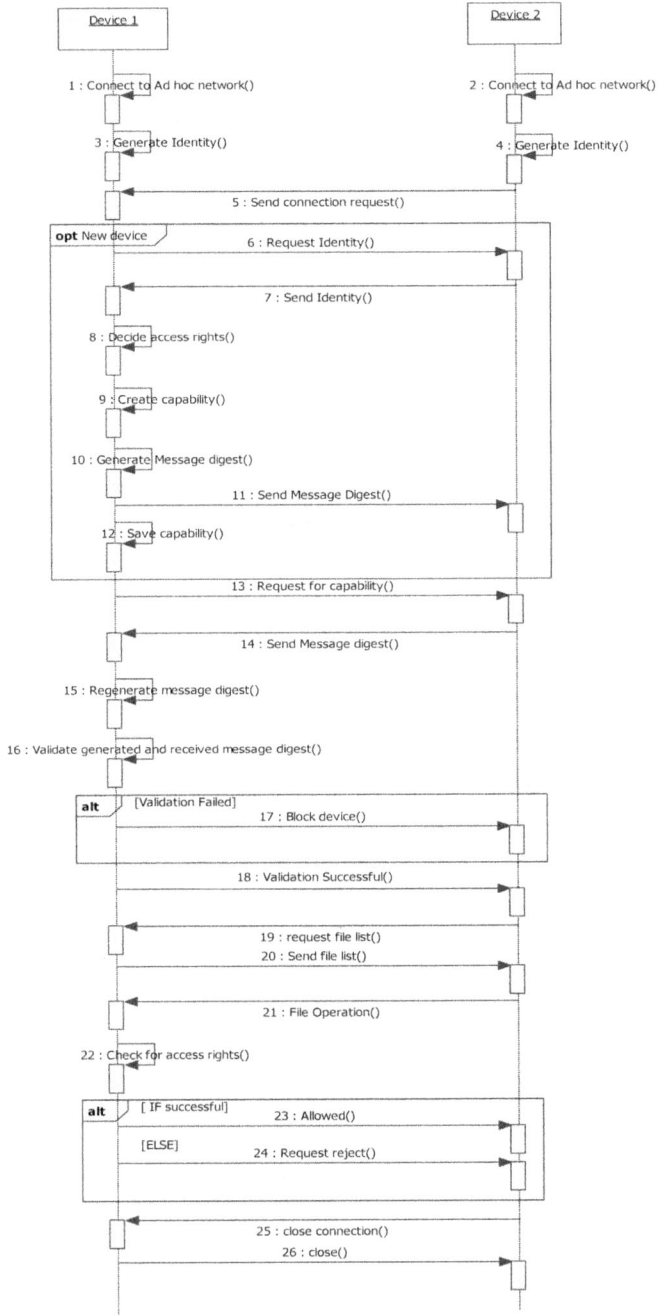

Figure 8 Proposed CAC scheme for IoT.

- ID: Device identifier,
- $f(\ )$: Knowledge of one way hash function.

6.1.2 Evaluation Results

The goal of evaluation is to verify protocol for attacks mentioned above and ensures mutual authentication along with the access control.

Mutual authentication: X_{uh} is shared securely between D_1 and D_2, and r is provided by trusted KDC to both the devices. Consequently, D_1 is authenticated to D_2 as only D_2 can decrypt R and T_{us}. Also MAC can be calculated only by D_2 and D_2 is sending the encrypted r to authenticate it to D_1. Verification results show that secure mutual authentication is achieved.

Man-in-the-middle attack: In case of authentication, even there is a man-in-the-middle attack on R, T_{us}, MAC_1 parameters; the attacker will not reveal any information. AVISPA shows that authentication protocol is free from attacks. For access control, man-in-the-middle attacks happen when an attacker eavesdrop the ID and ICAP transmitted, and then a masquerade attack happens when the attacker uses the stolen ID and CAP. The key to preventing a masquerade attack from the stolen CAP is to use an ID to validate the correct device. If the attacker manages to steal the ID, the attack is prevented by applying public key cryptography to ID, assuming that the authentication process has been done before access control. In this way, although the attacker gets the ICAP which is not encrypted, the capability validity check will return an exception because the one way hash function, $f(\text{ID, AR, Rnd})$ will return a different result than the one presented in the CAP, without a correct ID.

Another type of man-in-the-middle attack is replay attack. Adversary can intercept the message sent out from D_1. However, it is not possible in IACAC because it can easily detect by verifying timestamp T_u. If T_u is older than the predefined threshold value, it is invalid and has been used. If T_u is changed, $MAC_1 = MAC(X_{uh}, R\|ICAP_1)$ is not valid and consistent. For access control, IACAC prevents the replay attack by maintaining the freshness of Rnd, for example by using timestamp or nonce by including MAC as well. Even if the attacker manages to compromise the solution and gets the ICAP, it cannot use the same capability next time because the validity will be expired.

DoS attack: Upon receiving the message from D_1, D_2 first checks the validity of the timestamp. If it is not valid, then D_2 discards the message. Otherwise, it computes a MAC_2 value to compare with the received value. DoS happens when an attacker accesses a particular resource massively and simultaneously by using the same or different IDs. It is easy to control access using one ID because the system is able to maintain the session, thus the

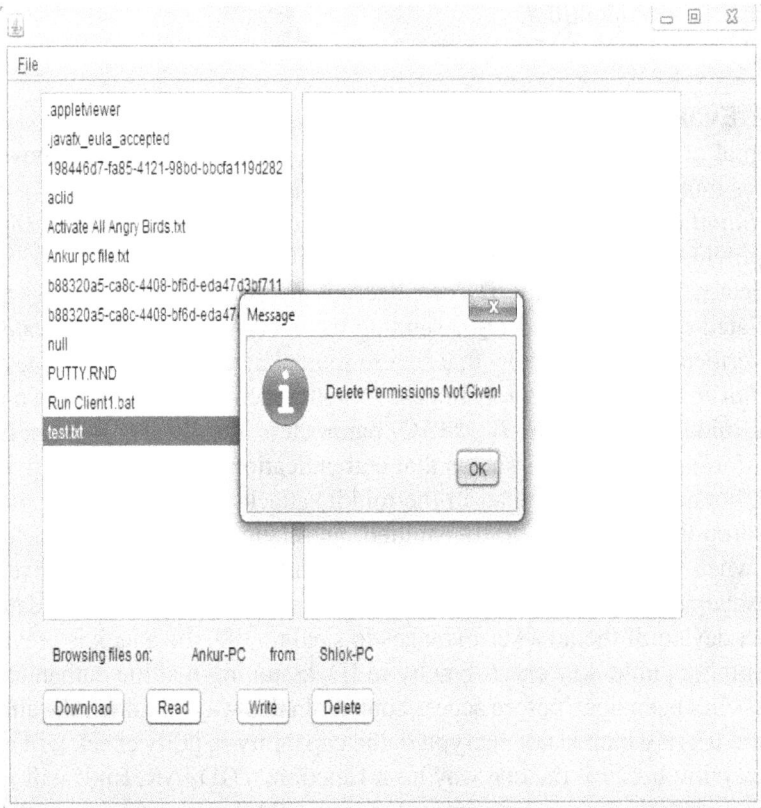

Figure 9 Snapshot showing principle of least privilege.

access of the same ID to the same resource can be restricted to only one session at a time. The potential of DoS attacks from multiple IDs can be prevented in the capability propagation process. Therefore, a DoS attack can be prevented or at least minimized.

Principle of Least Privilege: Security analysis shows that CAC has greater support for principle of least privilege due to the use of capabilities and hence it limits the damage when the protection is partially compromised. As access rights are encapsulated in the process of capability creation, even attacker or intruder is trying to modify these access rights, capability verification and comparison process returns false and access is denied. Access control schemes purely based on the role, context and ACL [44] has not addressed the principle of least privilege which is an important feature of the access control solution. A sample snapshot as in Figure 9 shows that even one device is trying to perform delete operation which is not included in its capability, delete operation is denied achieving the principle of least privilege.

6.2 Performance Analysis

6.2.1 IACAC

The security level of protocol presented in this paper depends on the type of MAC algorithm, encryption algorithm and security level of ECC signature. We propose to use RC5 stream cipher for encryption, which takes 0.26 ms on Mica2 motes [54–56]. RC5 is notable for its simplicity for resource constrained devices such as IoT and its flexibility due to the built in variability. Heavy use of data independent rotations and mixture of different operations provides strong security to RC5 [57]. We propose to use SHA-1 as one way hash function which takes 3.63 ms on Mica2 motes and it is computationally expensive to find text which matches given hash and also it is difficult to two different texts which produces the same hash [54–56]. To generate the MAC value, we propose CBC-MAC which has advantage of small key size and small number of block cipher invocations and takes 3.12 ms on Mica2 motes [55]. The time required to generate random number is 0.44 ms and ECC to perform point multiplication which takes 800 ms on Mica2 motes [55, 56]. In IACAC protocol as the message length is fixed, CBC-MAC is most secure [58]. It is clear from these values that maximum time is required for ECC point multiplication. In IACAC, point multiplication is taking place at KDC and as KDC is powerful device, computational overhead is trivial as compared to the sensors. We denote the computational time required for each operation by device in IoT by following notation:

- D_H is the time to perform one way hash function SHA-1;
- D_{MAC} is the time to generate Mac value by CBC-MAC;
- D_{RC5} is the time to perform encryption and decryption by RC5;
- D_{MUL} is the time to perform ECC point multiplication; and
- R is the time for random number generation.

Table 3 shows the comparison of computational time for the above-mentioned protocol. The IACAC protocol for mutual authentication and access control for the IoT devices takes less time (14.28 ms) as compared to other protocol compared in this paper. Key point to note here is that none of the work has addressed the issue of authentication and access control as an integrated solution for IoT. Total computational time for of the proposed

Table 3   Computational time for an IACAC scheme.

Scheme	IACAC	HBQ [59]	IoT_Auth [16]
Auth. Time	$2D_H + 2D_{MAC} + 2D_{RC5}$	$2D_H + 2D_{MAC} + D_{RC5} + 3D_{MUL}$	$R + D_H + 2D_{MUL}$
Total	$2D_H + 2D_{MAC} + 2D_{RC5}$	$2D_H + 2D_{MAC} + D_{RC5} + 3D_{MUL}$	$R + D_H + 2D_{MUL}$
Total time	14.02 ms	2413.76ms	1604.07ms

scheme, HBQ [59] and mutual authentication for IoT (IoT_Auth) [16] is shown in Table 3. IoT_Auth scheme requires $R + D_H + 2D_{MUL}$ time for mutual authentication which comes approximately 1604.07 ms. The HBQ scheme takes $2D_H + 2D_{MAC} + D_{RC5} + 3D_{MUL}$ total time for authentication which is approximately 2,413.76 ms. Key point to note here is that both schemes do not address access control after authentication. IACAC takes only $D_H + 2D_{MAC} + 2D_{RC5}$ which takes only 14.02 ms which is much better than the other two schemes analyzed in this paper. In IACAC, the $2D_H$ factor is introduced which comprises time required by one way hash function in authentication as well as in ICAP to calculate Rnd.

6.2.2 CAC

The performances of independent CAC have also been analyzed to validate and support our findings. The CAC implementation consists of the capability creation, object selection once capabilities are verified and denying access if there no match found for capability. In this paper, files are treated as objects and operations are performed as mentioned in capabilities. Operations are Read, Write, Read and Write, or NULL operations as explained earlier.

As stated earlier, the CAC scheme is implemented in Wi-Fi for Laptop devices. To check the performance of CAC in terms of Access Time (AT), different laptop devices of same configuration are used and AT is averaged for all devices. In this paper, AT is a function of latency and is defined as

$$\text{Access Time (AT)} = f(L) \tag{3}$$

where L is latency of access and defined as an overhead in terms of computational time to access right resource on right device. The unit of AT is milliseconds (ms). For measurement, we took the scenario as the two devices (Laptops) are connected via access point. AT defined in equation (3) is the time required to access one device to other in one way. Since WLAN is used and traffic can affect the access delay, multiple measurements are required

Table 4 Performance comparison of AT.

Scheme ⟶	CAC	CRBAC[44]
AT in (ms)	364	410

to consider for evaluation. The three measurement runs have been taken for calculating the access time. Two devices are discoverable to each other by the Jgroups [60]. JGroups is a reliable group communication toolkit implemented in Java. It is based on IP multicast, and also provide reliable group membership, lossless transmission of a message to all recipients, message ordering. As reliability requirement varies from application to application, JGroups provides a flexible protocol stack architecture that gives flexibility to users to put together custom-tailored stacks, ranging from unreliable but fast to highly reliable but slower stacks. There are two cases for the performance measure, first is access with capability and second without using capability. In both cases we considered the same common modules, as device discovery and file browsing.

Table 4 shows the performance comparison of CAC, AT without capability and CRBAC [44]. In this paper, we also implemented CRBAC scheme to check its performance with CAC scheme presented. In [44], programming framework is presented to model CRBAC. Same programming framework is implemented in Wi-Fi to get context aware role based access control for laptop devices. As per the framework presented in the paper, context management and access control are brought and implemented together to get role based access control. Performance in terms of AT in milliseconds (ms) is measured for CRBAC [44] access control scheme and it shows that CAC works better as compared to CRBAC. CAC take average AT of 364 ms and AT without capability take 173 ms. Table 4 shows that the CAC scheme takes extra 191 ms but it provides secure access to devices by avoiding tampering or forgery of capability with the help of one way hash function. CAC access is also attack resistant from replay and man-in-the-middle attack. The CRBAC scheme takes 410 ms to access the device, which is more than the CAC scheme. In the CRBAC context dependent role based access is granted but the access is not secure. It can be concluded from Table 4 that the CAC scheme gives secure access control with better performance in terms of AT.

Moreover, in a distributed context, like IoT, CAC provides many advantages over traditional or consolidated approaches due to its flexibility, better support for least privilege principle and avoidance for replay attack and man-in-the-middle attack. The chosen approach for the access control based on the capability concept, and in particular the CAC scheme, is considered in order to cope with the scalability of IoT system since it is well suited for providing access control in distributed systems. Besides a proposed access control model which provides scalability and flexibility, the main contribution of this paper also includes a secure access control mechanism that have been tested with a security protocol verification tool. To provide complete security solution to the identity management in IoT, authentication and access control are two important security measures.

Furthermore, there are few challenges to implement IACAC in mobile environment. Access delegation method with security considerations based on capability based context aware access control scheme intended for federated IoT networks is presented in [61]. In [61], capability propagation incorporating context in federated IoT environment with scalability and flexibility for distributed systems is presented. Authority delegation for mobile and federated environments is challenging due to dynamic and distributed nature. Another issue is that, it is necessary to have an established trust relationship between all entities prior to delegation. IACAC is completely compatible with the state of the art and it has been tested in Wi-Fi environment as discussed in the evaluation part of this paper. As the IACAC is addressing device to device authentication and access control, it is compatible in the user equipment and network elements being a lightweight and flexible in nature. Backward compatibility with the legacy network should not be the issue with the availability of high and powerful resources. In a mobile environment, mobility management is an interesting issue to deal with. The A interface which is an interface between mobile switching service switching centre and base station system which support many application part and Direct Transfer Application Part (DTAP) is one of them. Mobility management is one of the functionality of DTAP. There are many mobility management messages which are exchanged for identity establishment and access control (AUTHENT_REJ, AUTHENT_REQ). As physical layer of the A interface is 2 Mbps digital connection and DTAP deals with the exchange of layer 3 messages, no major adaptations are required to make IACAC functional.

As presented in [62], wireless communication and evolution is being faced by many constraints. These constraints are regulatory constraints like operating rules on the communication device, pre-decision on the frequency bands. Layered design of the communication protocol introduces architectural constraints which is important for proliferation. Other constraints are

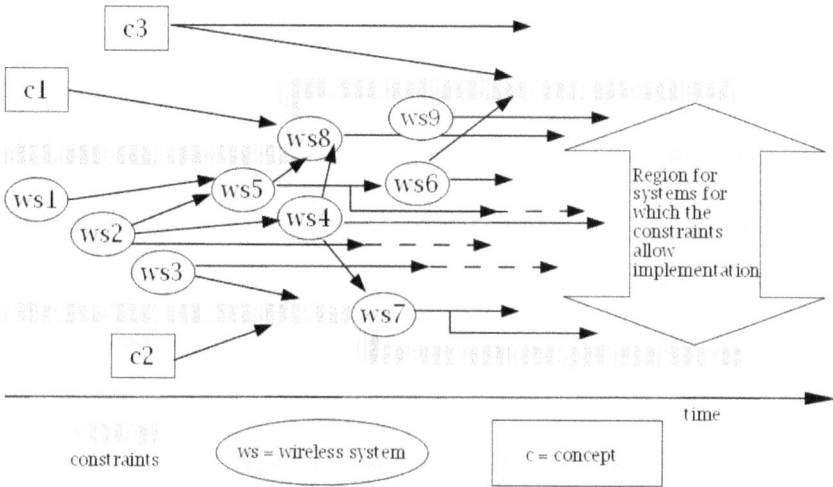

Figure 10 Wireless System Evolution [62].

standardization constraints in which particular communication protocol is developed and operated. The backward compatibility also needs many refinements and technological improvements for new standards. There are also market and social constraints deals with the new applications and the requirements from communication systems. Figure 9 depicts the outline of the evolution in wireless communications. As shown in the figure, ws1 and ws2 get converged and system ws5 is emerged. When ws4 is evolved, it is not feasible to implement concept c2 due to heavy constraints as discussed above, but due to increasing requirements (by ws3 also) the constraints are refined to change and ws7 is evolved. Over the period of time, some of the wireless communication systems become obsolete. Example of this obsolete system is shown in the Figure this happens for ws2. Important point to make a note here is that the constraints do not allow the concept c3 to be implemented over the period of time frame as depicted in Figure 10.

Similar to a global Internet scenario, interoperability and Internet working is ensured by following OSI stack but still there are many exceptions due to unpredictable nature of wireless interface. This makes more difficult to guarantee expected quality of service in resource constrained IoT and next generation networks. Backward compatibility to legacy networks is a challenge due to lack of cross layer coordination which is a need of today in order to get performance improvement. Other interoperability and Internet working

Queuing Model

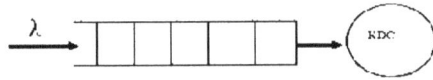

Figure 11 IACAC queuing model.

issues are architecture design and multi-traffic environment. To address these ensuing issues, more research is needed.

6.2.3 Proposed Mathematical Model for IACAC Queuing Analysis

The proposed IACAC model consists of a trusted third party which is responsible for distributing the ECC parameters to devices trying to communicate to each other. Devices approaching to KDC for service are managed in queue. Figure 11 shows the system, where λ is the arrival rate of devices. The inter-arrival time for devices is exponentially distributed. Thus arrival rate follows the poisons arrival process. Our system can be modelled with an M/D/1 queuing model with a constant service rate and one server. To evaluate the system performance, we model the sojourn time, that is, the total time spent by the device in the system.

The expectation of waiting time for devices in the queue can be as

$$E[W_q] = N_q \times E[S] + E[R] \tag{4}$$

where N_q is the mean number of devices in queue; $E[S]$ is the service time of KDC; and $E[R]$ is the residual time. Thus using Little's formula [63], the mean queue length is given as

$$N_q = \lambda \cdot E[W_q] \tag{5}$$

Therefore,

$$E[W_q] = \frac{E[R]}{1 - \rho_{KDC}}$$

where the utilization of KDC is given as

$$\rho_{KDC} = \lambda \cdot E[S]$$

The residual time R_i is the service time remaining to the customer being served when the ith device arrives at queue. Figure 12 shows the residual time in queue at time t.

R (t)

$\bar{R}$

S1   S2   Sn   t

Figure 12   Residual time in queue.

The mean residual time can be calculated by dividing the sum of areas of triangles by the length of interval and is derived as follows:

$$E[R] \;=\; \frac{1}{t}\int_0^t R(t)dt = \frac{1}{t}\sum_{i=1}^{n}\frac{1}{2}[S_i^2]$$

$$=\; \frac{n}{t}\cdot\frac{1}{n}\sum_{i=1}^{n}\frac{1}{2}[S_i^2]$$

$$\frac{n}{t}\to\lambda\quad \sum_{i=1}^{n}\frac{1}{2}[S_i^2]\to\frac{1}{2}E[S^2]$$

$$E[R] \;=\; \frac{\lambda\cdot E[S^2]}{2}$$

$$E[W_q] \;=\; \frac{\lambda\cdot E[S^2]}{2(1-\rho_{KDC})} \tag{6}$$

Now, the total time spent by a device in the system (the sojourn time) is

$$E[T] = E[W_q] + E[S]$$

$$E[T] = \frac{\lambda\cdot E[S^2]}{2(1-\rho_{KDC})} + E[S] \tag{7}$$

The total service time comprises of two factors: expectation $E[S]$ and variance $V[S]$. The variance is the difference between the mean of squares of the values and square of mean of values. Therefore $V[S]$ is given as

$$V[S] = E[S^2] - E[S]^2 \tag{8}$$

For the M/D/1 system, as the service time is constant, variance $V[S] = 0$ and results into $E[S^2] = E[S]^2$. Thus,

$$E[T] = \frac{\lambda \cdot E[S]^2}{2(1 - \rho_{KDC})} + E[S]$$

$$E[T] = \left(1 + \frac{\rho_{KDC}}{2(1 - \rho_{KDC})}\right) \cdot E[S] \tag{9}$$

By Little's formula, the mean queue length, the mean number of devices in queue is given by

$$N_q = \lambda \cdot E[W_q]$$

$$N_q = \frac{\lambda 2 \cdot E[S]^2}{2(1 - \rho_{KDC})}$$

$$N_q = \frac{\rho_{KDC}^2}{2(1 - \rho_{KDC})} \tag{10}$$

Thus, from equations (4) to (10), it can be concluded that the total time spent by a device in system is the function of the service time $E[S]$ and the utilization of KDC, ρ_{KDC}. The mean queue length and the utilization are proportional to each other. If the number of devices in queue increases, the utilization of KDC also increases. For further improvement in the utilization of KDC, we can pipeline the services of KDC. The services provided by KDC can be divided in three stages. This will lead to service of three devices at a time. As shown in Figure 13, the server device will get serviced from server S1 and will enter the queue for server S2 and so on.

Thus a network of set of single servers in series is formed. The input for each queue except for the first is the output of the previous queue. The input to the first queue is Poisson. If the service time of each queue is constant and the waiting lines are infinite, the output of each queue is a Poisson stream statistically identical to the input. When this stream is fed into the next queue, the delays at the second queue are the same as if the original traffic had bypassed the first queue and fed directly into the second queue. Thus the queues are independent and may be analysed one at a time. Therefore the

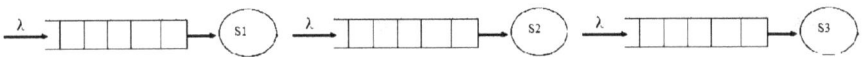

Figure 13 Proposed pipelining of the KDC services.

waiting time for a device in complete system will be the sum of waiting time for devices at each subsystem and is shown as

$$T = \sum E[T_i]$$

$$T = \sum_{i=1}^{3} \left(1 + \frac{\rho_i}{2(1 - \rho_i)}\right) \cdot E[S_i] \tag{11}$$

where ρ_i is the utilization of server S_i and $E[S_i]$ is the service time of server S_i.

7 Conclusions and Future Work

A distributed, lightweight and attack resistant solution are the mandatory properties for the security solution in IoT and puts resilient challenges for authentication and access control of devices. This paper presents an efficient and secure ECC based integrated authentication and access control protocol. This paper also presents a mutual authentication protocol and integrated with novel and secure approach of CAC for access control in IoT along with the implementation results. Furthermore, this paper presents comparative analysis of different authentication and access control schemes for IoT. Comparison in terms of computational time shows that IACAC scheme is efficient as compared to other solution. The protocol is also analyzed for the performance and security point of view for different possible attacks in IoT scenario. Protocol evaluation shows that it can defy attacks like DoS, man-in-the-middle and replay attacks efficiently and effectively. The paper also presents protocol verification using AVISPA tool which proves that the IACAC protocol is also efficient in terms of key sharing and authentication. Finally, we also presented a mathematical model for improving queuing analysis of IACAC.

The future plan is to put this protocol in place with RFID middleware architecture for identity management in IoT. Future work will involve specification as well as security evaluation of the CAC propagation and revocation in order to have a complete model of CAC scheme. Another interesting aspect will be to define and devise a lightweight version of CAC for resource constrained devices in IoT like sensor nodes. Complete interoperability and Internet working is still an open research area to take this research further.

References

[1] ITU-T Internet Reports, Internet of Things, November 2005.
[2] E. Zouganeli and I. E. Svinnset. Connected objects and the Internet of Things – A paradigm shift, Photonics in Switching 2009, September 2009.
[3] M. Weiser, The computer for the 21st century, Scientific American, 265: 66–75, 1991.
[4] S. Sarma, D. L. Brock, and K. Ashton. The networked physical world. TR MIT-AUTOIDWH-001, MIT Auto-ID Center, 2000.
[5] Jayavardhana Gubbi, Rajkumar Buyya, Slaven Marusic, and Marimuthu Palaniswami. Internet of Things (IoT): A vision, architectural elements, and future directions. Technical Report CLOUDS-TR-2012-2, Cloud Computing and Distributed Systems Laboratory, The University of Melbourne, 29 June 2012.
[6] Xiaodong Lin, Rongxing Lu, Xuemin Shen, Y. Nemoto, and N. Kato. Sage: A strong privacy-preserving scheme against global eavesdropping for ehealth systems. IEEE Journal on Selected Areas in Communications, 27(4): 365–378, May 2009.
[7] A. Gluhak, S. Krco, M. Nati, D. Pfisterer, N. Mitton, and T. Razafindralambo. A survey on facilities for experimental Internet of Things Research. IEEE Commun. Mag., 49: 58–67, 2011.
[8] P. Spiess, S. Karnouskos, D. Guinard, D. Savio, O. Baecker, L. Souza, and V. Trifa. SOA-based integration of the internet of things in enterprise services. In Proceedings of IEEE ICWS 2009, Los Angeles, Ca, USA, July 2009.
[9] I. F. Akyildiz, J. Xie, and S. Mohanty. A survey on mobility management in next generation All-IP based wireless systems. IEEE Wireless Communications Magazine, 11(4):16–28, 2004.
[10] C. Mayer. Security and privacy challenges in the IoT. WowKivs, Electronic Communications of the EASST, Volume 17, Germany, 2009.
[11] R. Prasad. My personal Adaptive Global NET (MAGNET). Signals and Communication Technology Book, Springer, The Netherlands, 2010.
[12] D. M. Kyriazanos, G. I. Stassinopoulos, and N. R. Prasad. Ubiquitous access control and policy management in personal networks. In Third Annual International Conference on Mobile and Ubiquitous Systems: Networking & Services, pp. 1–6, July 2006.
[13] Michael Braun, Erwin Hess, and Bernd Meyer. Using elliptic curves on RFID tags. International Journal of Computer Science and Network Security, 8(2), 2008.
[14] Sheikh Iqbal Ahamed, Farzana Rahman, and Endadul Hoque. ERAP: ECC based RFID authentication protocol. In 12th IEEE International Workshop on Future Trends of Distributed Computing Systems, 2008.
[15] D. Balfanz, D. K. Smetters, P. Stewart, and H. C. Wong. Talking to strangers: Authentication in ad-hoc wireless networks. In Network and Distributed Systems Security Symposium (NDSS), February 2002.
[16] Guanglei Zhao, Xianping Si, Jingcheng Wang, Xiao Long, and Ting Hu. A novel mutual authentication scheme for Internet of Things. In Proceedings of 2011 IEEE International Conference on Modelling, Identification and Control (ICMIC), pp. 563–566, 26–29 June 2011.
[17] C. Jiang, B. Li, and H. Xu. An efficient scheme for user authentication in wireless sensor networks. In 21st International Conference on Advanced Information Networking and Applications Workshops, pp. 438–442, 2007.

[18] R. R. S. Verma, D. O'Mahony, and H. Tewari. Progressive authentication in ad hoc networks. In Proceedings of the Fifth European Wireless Conference, February 2004.
[19] T. Suen and A. Yasinsac. Ad hoc network security: Peer identification and authentication using signal properties. In Proceedings from the Sixth Annual IEEE SMC Information Assurance Workshop (IAW'05), pp. 432–433, 15–17 June 2005.
[20] L. Venkatraman and D. P. Agrawal. A novel authentication scheme for ad hoc networks. In Wireless Communications and Networking Conference (WCNC2000), vol.3, pp. 1268–1273. IEEE, 2000.
[21] B. Bing. Emerging Technologies in Wireless LANs – Theory, Design and Deployment. Cambridge University Press, 2008.
[22] Best Current Practices for WISP Roaming, WiFi Alliance, 2003.
[23] RFC 2865, Remote Authentication Dial in User Service (RADIUS).
[24] Jian Feng. Analysis, implementation and extensions of RADIUS protocol. In International Conference on Networking and Digital Society (ICNDS'09), vol.1, pp. 154–157, 30–31 May 2009.
[25] RFC 5247, Extensible Authentication Protocol (EAP) Key Management Framework, August 2008.
[26] A. M. El-Nagar, A. A. El-Hafez, and A. Elhnawy. A novel EAP – Moderate weight Extensible Authentication Protocol. In IEEE Seventh International Conference on Computer Engineering (ICENCO2011), pp. -1-6, 27–28 December 2011.
[27] Wei Yuan, Liang Hu, Hong-tu Li, Kuo Zhao, Jiang-feng Chu, and Yuyu Sun. Key replicating attack on an identity-based three-party authenticated key agreement protocol. In IEEE International Conference on Network Computing and Information Security (NCIS), vol. 2, pp. 249–253, 14–15 May 2011.
[28] Jun Lei, Xiaoming Fu, Dieter Hogrefe, and Jianrong Tan. Comparative studies on authentication and key exchange methods for 802.11 wireless LAN. Computers & Security, 26(5): 401–409, August 2007.
[29] OASIS.eXtensible Access Control Markup Language (XACML) Version 3.0, Working Draft 8, February 2009.
[30] W3C Platform for Privacy Project: http://www.w3.org/privacy/.
[31] The Shibboleth project: www.shibboleth.net.
[32] The Liberty Alliance Project: www.projectliberty.org.
[33] Ravi S. Sandhu. The typed access matrix model. In Proceedings of the IEEE Symposium on Security and Privacy. IEEE CS Press, 1992.
[34] T. Close. ACLs don't. HP Laboratories Technical Report, February 2009.
[35] L. Gong. A secure identity-based capability system. In Proceedings of 1989 IEEE Symposium on Security and Privacy, Oakland, CA, May. IEEE Computer Society Press, Los Alamitos, 1989.
[36] Ravi S. Sandhu, E. J. Coyne, H. L. Feinstein, and C. E. Youman. Role-based access control models. IEEE Computer, 29(2): 38–47, February 1996.
[37] J. B. D. Joshi, E. Bertino, U.Latif, and A. Ghafoor. A generalized temporal role-based access control model. IEEE Transactions on Knowledge and Data Engineering, 17(1): 4–23, January 2005.
[38] R. Bhatti, E. Bertino, and A. Ghafoor. A trust-based context-aware access control model for web-services. Distributed and Parallel Databases, 18(1), July 2005.

[39] Q. Ni, A. Trombetta, E. Bertino, and J. Lobo. Privacy-aware role based access control. In Proceedings of the 12th ACM Symposium on Access Control Models and Technologies (SACMAT'07), 2007.

[40] E. Barka and R. Sandhu. A role-based delegation model and some extensions. In Proceedings of the 23rd National Information Systems Security Conference, 2000.

[41] E. Barka and R. Sandhu. Role-based delegation model/hierarchical roles. In Proceedings of the 20th Annual Computer Security Applications Conference (ACSAC'04), 2004.

[42] K. Hasebe, M. Mabuchi, and A. Matsushita. Capability-based delegation model in RBAC. In Proceedings of the 15th ACM Symposium on Access Control Models and Technologies (SACMAT'10). ACM, 2010.

[43] Y. G. Kim, C. J. Mon, D. Jeong, J. O. Lee, C. Y. Song, and D. K. Baik. Context-aware access control mechanism for ubiquitous applications. In Advances in Web Intelligence, LNCS, Vol. 3528, pp. 236–242. Springer, Heidelberg, 2005.

[44] D. Kulkarni and A. Tripathi. Context-aware role-based access control in pervasive computing systems. In SACMAT'08, Estes Park, CO, 11–13 June 2008.

[45] Kyungah Shim and Young-Ran Lee. Security flaws in authentication and key establishment protocols for mobile communications. Applied Mathematics and Computation, 169(1): 62–74, October 2005.

[46] G. Horn, K. M. Martin, and C. J. Mitchell. Authentication protocols for mobile network environment value added services. IEEE Transactions on Vehicular Technology, 51(2): 383–392, 2002.

[47] M. J. Beller, L. F. Chang, and Y. Yacobi. Privacy and authentication on a portable communications system. IEEE Journal on Selected Areas in Communications, 11: 821–829, 1993.

[48] C. Boyd and A. Mathuria. Key establishment protocols for Secure Mobile communications: A selective survey. In Information Security and Privacy, ACISP 98, LNCS, Vol. 1438, pp. 344–355. Springer, Heidelberg, 1998.

[49] A. Aziz and W. Diffie. Privacy and authentication for wireless local area networks. IEEE Personal Communications, 1: 25–31, 1994.

[50] D. S. Wong and A. H. Chan. Efficient and mutually authenticated key exchange for low power mobile device. In Advances in Cryptology – Asiarcypt01, LNCS, Vol. 2248, pp. 272–289. Springer-Verlag, Heidelberg, 2001.

[51] N. Koblitz. Elliptic curve cryptosystems. Mathematics of Computation, 48: 203–209, 1987.

[52] Avispa – A tool for Automated Validation of Internet Security Protocols. http://www.avispa-project.org.

[53] D. Dolev and A. C.-C. Yao. On the security of public key protocols. In FOCS, pp. 350–357. IEEE, 1981.

[54] R. Chakravorty. A programmable service architecture for mobile medical care. In 4th IEEE International Conference on Pervasive Computing and Communications, 2006.

[55] C. Karlof, N. Sastry, and D. Wagner. Tinysec: Link layer security architecture for wireless sensor networks. In SenSys, ACM Conference on Embedded Networked Sensor Systems, 2004.

[56] N. Gura, A. Patel, A. Wander, H. Eberle, and S. C. Shantz. Comparing elliptic curve cryptography and RSA on 8-it CPUs. In CHES 2004, LNCS, Vol. 3156, pp. 119–132, Springer, Heidelberg, 2004.

[57] Y. L. Yin. The RC5 encryption algorithm: Two years on. CryptoBytes, 3(2), Winter 1997.

[58] M. Bellare, J. Killan, and P. Rogaway. The security of cipher block chaining. In Y. Desmedt (Ed.), CRYPTO 1994. LNCS, Vol. 839, pp. 341–358. Springer, Heidelberg, 1994.

[59] H. Wang, B. Sheng, and Q. Li. Elliptic curve cryptography based access control in sensor networks. Int. J. Security and Networks, 1(3/4): 127–137, 2006.

[60] Bela Ban. Adding group communication to Java in a non-intrusive way using the ensemble toolkit. Technical Report, Dept. of Computer Science, Cornell University, November 1997.

[61] Bayu Anggorojati, Parikshit N. Mahalle, Neeli R. Prasad, and Ramjee Prasad. Capability-based access control delegation model on the federated IoT network. In IEEE 15th International Symposium on Wireless Personal Multimedia Communications (WPMC2012), Taipei, Taiwan, September 24–27, pp. 604–608, 2012.

[62] Petar Popovski. On designing future communication systems: Some clean-slate perspectives. In R. Prasad, S. Dixit, R. Nee, and T. Ojanpera (Eds.), Globalization of Mobile and Wireless Communications, pp. 129–143. Springer Science+Business Media, 2011.

[63] Alberto Leon-Garcia. Probability, Statistics, and Random Processes for Electrical Engineering (3rd ed.). Prentice Hall, 2008.

Biographies

Parikshit N. Mahalle is IEEE member, ACM member, Life member ISTE and graduated in Computer Engineering from Amravati University, Maharashtra, India in 2000 and received Master in Computer Engineering from Pune University in 2007. From 2000 to 2005, he was working as lecturer in Vishwakarma Institute of technology, Pune, India. From August 2005, he was working as an Assistant Professor in Department of Computer Engineering, STES's Smt. Kashibai Navale College of Engineering, and Pune, India. Currently he is pursuing his Ph.D. in wireless communication at Center for TeleInFrastruktur (CTIF), Aalborg University, Denmark. He has published 25 papers at national and international level. He has authored five books on subjects like Data Structures, Theory of Computations and Programming Languages. He is also the recipient of "Best Faculty Award" by STES and Cognizant Technologies Solutions. His research interests are Algorithms, IoT, Identity Management and Security.

Bayu Anggorojati is currently pursuing his PhD at Center for TeleIn-Frastruktur (CTIF), Aalborg University. His main research interest is in access control for RFID system and IoT. During the period of his PhD work, he has been involved in several projects, especially the EC projects, such as ASPIRE, ISISEMD, LIFE2.0, and BETaaS.

Neeli Rashmi Prasad, Ph.D., IEEE Senior Member, Director, Center For TeleInfrastructure USA (CTIF-USA), Princeton, USA. She is also Head of Research and Coordinator of Themantic area Network without Borders,

Center for TeleInfrastruktur (CTIF) headoffice, Aalborg University, Aalborg, Denmark.

She is leading IoT Testbed at Easy Life Lab (IoT/M2M and eHealth) and Secure Cognitive radio network testbed at S-Cogito Lab (Network Management, Security, Planning , etc.). She received her Ph.D. from University of Rome "Tor Vergata", Rome, Italy, in the field of "adaptive security for wireless heterogeneous networks" in 2004 and M.Sc. (Ir.) degree in Electrical Engineering from Delft University of Technology, the Netherlands, in the field of "Indoor Wireless Communications using Slotted ISMA Protocols" in 1997.

She has over 15 years of management and research experience both in industry and academia. She has gained a large and strong experience into the administrative and project coordination of EU-funded and Industrial research projects. She joined Libertel (now Vodafone NL), The Netherlands in 1997. Until May 2001, she worked at Wireless LANs in Wireless Communications and Networking Division of Lucent Technologie, the Netherlands. From June 2001 to July 2003, she was with T-Mobile Netherlands, the Netherlands. Subsequently, from July 2003 to April 2004, at PCOM:I3, Aalborg, Denmark. She has been involved in a number of EU-funded R&D projects, including FP7 CP Betaas for M2M & Cloud, FP7 IP ISISEMD ICt for Demetia, FP7 IP ASPIRE RFID and Middleware, FP7 IP FUTON Wired-Wireless Convergence, FP6 IP eSENSE WSNs, FP6 NoE CRUISE WSNs, FP6 IP MAGNET and FP6 IP Magnet Beyond Secure Personal Networks/Future Internet as the latest ones. She is currently the project coordinator of the FP7 CIP-PSP LIFE 2.0 and IST IP ASPIRE and was project coordinator of FP6 NoE CRUISE. She was also the leader of EC Cluster for Mesh and Sensor Networks and is Counselor of IEEE Student Branch, Aalborg. Her current research interests are in the area of IoT & M2M, Cloud, identity management, mobility and network management; practical radio resource management; security, privacy and trust. Experience in other fields includes physical layer techniques, policy based management, short-range communications. She has published over 160 publications ranging from top journals, international conferences and chapters in books. She is and has been in the organization and TPC member of several international conferences. She is the co-editor is chief of *Journal for Cyber Security and Mobility* by River Publishers and associate editor of *Social Media and Social Networking* by Springer.

Ramjee Prasad (R) is currently the Director of the Center for TeleInfrastruktur (CTIF) at Aalborg University (AAU), Denmark and Professor, Wireless Information Multimedia Communication Chair. He is the Founding Chairman of the Global ICT Standardisation Forum for India (GISFI: www.gisfi.org) established in 2009. GISFI has the purpose of increasing the collaboration between European, Indian, Japanese, North-American, and other worldwide standardization activities in the area of Information and Communication Technology (ICT) and related application areas. He was the Founding Chairman of the HERMES Partnership – a network of leading independent European research centres established in 1997, of which he is now the Honorary Chair.

Ramjee Prasad is the founding editor-in-chief of the Springer *International Journal on Wireless Personal Communications*. He is a member of the editorial board of several other renowned international journals, including those of River Publishers. He is a member of the Steering, Advisory, and Technical Program committees of many renowned annual international conferences, including Wireless Personal Multimedia Communications Symposium (WPMC) and Wireless VITAE. He is a Fellow of the Institute of Electrical and Electronic Engineers (IEEE), USA, the Institution of Electronics and Telecommunications Engineers (IETE), India, the Institution of Engineering and Technology (IET), UK, and a member of the Netherlands Electronics and Radio Society (NERG) and the Danish Engineering Society (IDA). He is also a Knight ("Ridder") of the Order of Dannebrog (2010), a distinguishment awarded by the Queen of Denmark.

The Number Continuity Service: Part I – GSM <-> Satellite Phone

Arnaud Henry-Labordère

HALYS, Paris, France and PRISM-CNRS, Versailles, France; e-mail: ahl@halys.fr

Received 15 January 2013; Accepted 17 February 2013

Abstract

Mobile Number Portability is now a widely used service allowing users to keep their number if they change their subscribed operator. It was first deployed in Hong-Kong (1999). "Number Continuity" corresponds to the same service but with the *switch to another technology*, because the subscribed main terminal does not have coverage. The new terminal may be a PC, a smartphone/WiFi, a "satphone", a GSM phone if the main one is CDMA, with (almost) the same service transparently. Making calls or SMS with its normal CLI shown, receiving calls, SMS, MMS to his normal GSM number (unlike "Skype"). This article explains the GSM <-> satphone number continuity implementation. The Geostationary and Low Earth Orbit characteristics are presented as well the consequences of orbit drifts for which the exact computation is given based on the formal integration of Kepler's area law. The handover implementation and cases for the satellite service are explained. The constellation and services of the main satellite operators are compared. The telecom core network implementation of the service is detailed in the cases of a GSM type core network and an IS-41 (CDMA) core network. Subsequent articles will cover GSM<->CDMA (IS-41) and GSM<->WiFi.

Keywords: number continuity, satphone, IS-41, handover, satellite services, IMSI nominal, IMSI auxiliary, multi-IMSI, geostationary, LEO.

1 User Benefit from the GSM <-> Satellite Number Continuity Implementation

A customer owns two mobile phones: one GSM mobile and one satellite phone. His contacts know only his GSM number and he will receive few calls to his GSM numbers as no one knows he is using a satellite phone. He is currently in an uncovered GSM area, ship, or open country but his satellite phone is activated after he has turned it on. With the Number Continuity implemented for his satellite phone, he will receive all calls and SMS to his satellite phone even they were sent to his GSM number. The only thing he has done is turning on the satellite phone. No manual call forwarding (it would not forward the SMS) and if he has lost GSM coverage he could not use the call forwarding function. If he is again under GSM coverage, he reactivates it, and again all calls and SMS will be received by the GSM handset.

The Number Continuity Solution allows this transparent service for GSM and Satellite phone customers. It is implemented by satellite operators, or by Roaming Hub operators in cooperation with them and with agreements with the involved GSM operators.

2 Satellite Coverage and Position Station Keeping of GEO and LEO

2.1 Telecom Coverage

For GEO and LEO the orbit plane is supposed to be the equator (which makes an angle of 23°27' (the tilt of the earth axis) with the earth ecliptic plane, where all the other planets are mostly (Pluto has an inclined orbit). For a LEO satellite, the orbit is strongly inclined at the equator.

From Newton's gravitation law, we have:

- $\gamma_g = K/R^2$ for the terrestrial acceleration where R is the geostationary satellite *distance* from the earth center. The gravitation field decreases as the square of the distance to the earth center.
- $\gamma_c = \omega^2 \times R$ is the centrifugal acceleration applied to the satellite due to its circular assumed orbit around the center of an earth inertial coordinate system.

The earth (same angular speed exactly as satellite) has a rotation time in one sidereal day $D = 23$ h 56 m 4 s (not exactly the legal 24 hours), which gives the common angular speed ω of the earth and of the geostationary satellite:

$$\omega = 2\Pi/D = 0.000072921173 \text{ rad/sec}$$

K is such that $\gamma_g = 9.822$ m/sec^2 (the *standard gravity*) at $R_0 = 6371$ km the official geodetic radius of the "geoïd surface", hence:

$$K = 9.822 \times (R_0)^2 = 3.986714 \times 10^{14} \quad \text{(in m}^3\text{/sec}^2\text{)}$$

Setting the gravitation equal to the centrifugal acceleration $\gamma_g = \gamma_c$ for a circular orbit, with the above values of K and ω yields

$$R = (K/\omega^2)^{1/3} = 42166 \text{ km} \quad \text{(distance from the earth center)}$$

which gives an *altitude* of $R - R_0 = 35795$ km above the standard ground level if the orbit was circular, commonly approximated by 36000 km and it is independent of the mass of the satellite.

If the orbit was circular and the inclination of the orbit equal to 0° with the equator, the satellite would look still from a ground telecom station, which is simple for tracking.

The satellite covers the earth within a cone which 1/2 angle is

$$\arccos(R_0/R) = 81° \text{ latitude}$$

This is better for telecom coverage of the north zone that most LEO systems with inclined orbits.

2.2 West-East Drift (Deviation from a Theoretical Circular Orbit)

When the satellite is launched, the orbit is not exactly circular (this is a delicate and fuel consuming launch phase from an elliptical "transfer orbit") even if we assume that the revolution time is exactly one sidereal day. The satellite has an elliptical orbit (Kepler's ellipse law) with a small excentricity e ($e = 0$ for a circle, and $e < 0$ for an ellipse). Its equation in polar coordinates is

$$\rho = P/(1 + e\cos\theta)$$

P is the distance from the focus F_1 (the centre of the earth) to the ellipse. To test real numerical examples, here are the classical relations between the parameters of an ellipse in polar coordinates and in Cartesian coordinates:

$P = a(1 - e^2)$, easy to find, by computing ρ for $\theta = 0$ (apogee) and $\theta = \pi$ (perigee)

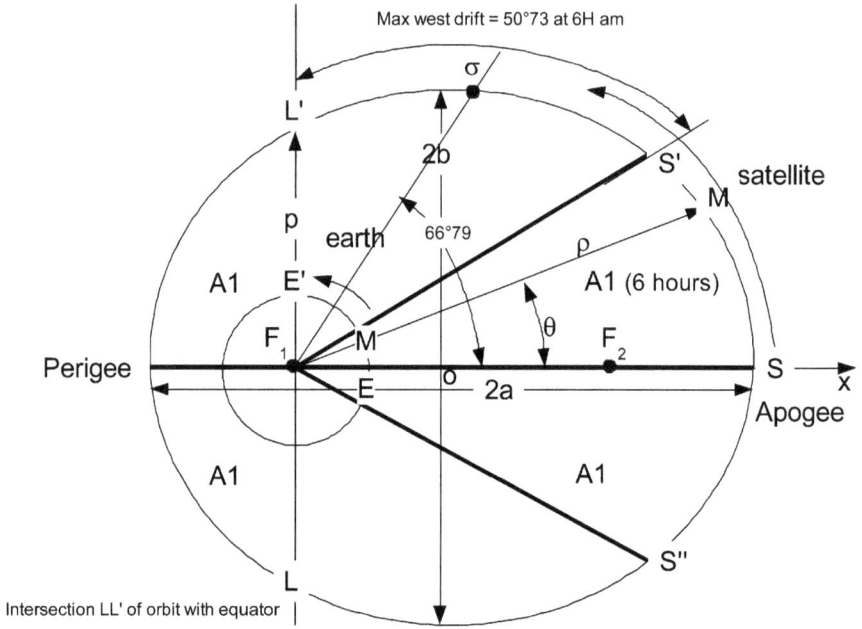

Max west drift = 50°73 at 6H am

Example of defective geostationnary orbit: excentricity too large = -0.507

Figure 1   Geostationary orbit with eccentricity $e \neq 0$.

$P = b^2/a$
$c = ae$ (1/2 distance between the two focus F_1 and F_2)
A (the area inside the ellipse) $= \pi ab$

The perigee (lowest altitude) is P and the apogee (highest) is A.

Kepler's law of equal areas says that an equal area of the ellipse is covered in a given time interval. Assume that at 0 h, the satellite is in S vertical of the earth E (the nominal satellite longitude). Six hours later, the satellite is in S' (the area $A_1 = AS'F_1$ is equal to the area $F_1S'P$ as P is the satellite position at 12 h). At 0, 12 and 24 h again, the satellite is at its nominal longitude. However, in the AS' part of the day, the satellite is revolving slower that the earth and appears to drift westward (the earth turns west to east). When the satellite is in S', the drift difference is the angle $E'F_1S'$. In the $S'P$ part, it is faster and appears to drift eastward. An earth observer sees the satellite drifting westward from 0 h to T and coming back to its nominal longitude from T to 12 h, drifting eastward from 12 h to $(24\,h - T)$ and coming back to its nominal longitude from $24\,h - T$ to 24 h.

If the orbit is far from being circular, the *E-W* telecommunication coverage may be defective during the two daily periods of maximum drift.

2.2.1 Angular Position $\theta(t)$ of the Satellite

The total area A of the ellipse can be computed simply as above knowing P and e. The differential of A for a small angle θ, is

$$dA = \frac{1}{2}P^2 d\theta \quad \text{(area of a small triangle with height } \rho \text{ and basis } \rho d\theta) \quad (1)$$

To get $\theta(t)$ we use Kepler's equal area law.

Assuming to simplify that the angular position of the satellite is $S = 0$ at 0 h we have

$$S\frac{1}{2}(P^2(1-e\cos\theta)^2)d\theta = (A/D)t \text{ the area of the ellipse from 0 to } t \text{ time)0, } t$$

Kepler's law: Area between time 0 and $T = \displaystyle\int_0^T \frac{P^2}{(1 + e\cos\theta)^2}d\theta = kT \quad (2)$

Taking $T = D$ (full revolution in one day 23 h 56 m 4 s), the total ellipse area being A,

$$A = kD \rightarrow k = \frac{A}{D} \quad (3)$$

This is the integral equation relating to time T and the angular position $\theta(T)$.

It is complicated to integrate by hand but this is a "definite integral" (using elementary functions), the online formal integration tool [3] gives the result using real valued functions only for $|e| > 1$ so the trick is to set the eccentricity $e' = 1/e$ in [2] and do instead a formal integration of

$$g(\theta) = \int \frac{1}{\left(1 + \left(\frac{1}{e}\right)\cos\theta\right)^2}d\theta \quad (4)$$

which gives an expression of real valued functions ($e^2 - 1$ is > 0):

$$e^2 \left(\frac{2e\tan^{-1}\left(\frac{(e-1)\tan\left(\frac{\theta}{2}\right)}{\sqrt{e^2-1}}\right)}{(e^2-1)^{3/2}} - \frac{\sin(\theta)}{(e^2-1)\cos(\theta)+e} \right) \quad (5)$$

Using Kepler's area law [2], the value of k in [3] and the expression of $g(\theta)$ [5] yields

$$g(\theta(T)) = \frac{A}{DP^2}T \qquad (6)$$

This gives

$$\theta(T) = g^{-1}\left(\frac{A}{DP^2}T\right) \qquad (7)$$

(but g^{-1} can only be computed numerically).

2.2.2 C Program to Compute the Time T to Have the Angle θ

(Thanks to Pascal Adjamagbo and Jean-Yves Charbonnel (Institut de Mathématiques de Jussieu (IMJ), November 2012.)

```
/*-------------------------------------------------------------------*/
/* Computes the time in hours for the satellite to go from 0 to
   theta                                                             */
/* on an elliptic orbit following Kepler's law                       */
/* AHL 22/11/2012                                                    */
/* Method: use the primitive obtained by formal integration, see [5] */
/* then applies Kepler's law to obtain the time                      */
/* ENTRY; theta: angle in radians                                    */
/*             eprime: eccentricity of ellipse -1 < e < 0(circle)    */
/*             P ellipse parameter in km                             */
/*             AS : 1/2 area of ellipse reached at day/2, in km2     */
/* constant : D (in 1/2 days, with day) is 23H 56min 4 sec           */
/* RESULT: TIME in hours                                             */
/*------------------------------------------------------------------ */
double Temps(double theta, double eprime, double P , double AS)
{
  double TIME, Prim;
  double e, e2moins1;

  e = 1/eprime; //inverse of eccentricity is then < -1 for an ellipse
  e2moins1 = e*e - 1; // which is > 0 !! used to simplify the
  expression of the primitive // result of formal integration to obtain
  the primitive Prim = (e*e)*(
          (2*e/(e2moins1*sqrt(e2moins1)))* atan( (e-1)*tan(theta/2)/
          sqrt(e2moins1) ) - sin(theta)/(e2moins1 * (cos(theta) + e))
          );
  TIME = D * (0.5 * P*P * Prim) / AS; //Kepler's law :time in hours
  from 0 to theta as a proportion of the 1/2 ellipse area AS
  return(TIME);
}
```

2.2.3 Computation of Angle of Radial Speed Synchronism

The radial speed of the satellite and of the earth are the same for an angular position θ_s such as

$$\frac{P^2}{(1 + e \cos \theta_s)^2} = \frac{A}{\pi}$$

that is,

$$\theta_s = \arccos \left(\frac{P\sqrt{\frac{\pi}{A}} - 1}{e} \right)$$

which we represent as σ in Figure 1.

2.2.4 Computation of the Maximum Westward and Eastward Daily Drift

The drift is the difference in longitude between the normal longitude of the geostationary satellite and the real longitude.

$\theta_s = 66°79'$ with the above numerical values. At this angular position, the west drift is maximum because the satellite becomes faster eastward than the earth. The numerical integration gives $t_s = 8\,\text{h}\,27\,\text{m}$. We can compute the earth angular position and the drift.

Normally, the tolerance given by the space organisations for geostationary station keeping is about 1° which means that if the launch does not succeed in setting an almost circular orbit, the satellite is not usable. Some telecom operators

2.2.5 Long-Term West or East Drift

The drift is due to the earth potential not being symmetrical (the equator is slightly elliptical), and has a tendency to "pull" the latitude toward two stable

Table 1   West-east and north-east drift depending on the orbit's eccentricity.

Excentricity	$S' =$ maximum westward drift at 6 h am	Duration of north drift $= g(\pi/2 - 0)$	Duration of south drift $= g(\pi - \pi/2)$	Angular position of synchronism
−0.507	60°08	9 h 42 m	2 h 18 m	66°79
...	...	...	...	...
−0.017	1°95	6 h 8 m	5 h 52 m	89°27
−0.012	1°38	6 h 5 m	5 h 55 m	89°48
−0.007	0°80	6 h 3 m	5 h 57 m	89°69
−0.002	0°23	6 h 1 m	5 h 59 m	89°91
0 (circle)	0°	6 h	6 h	All

"tesseral points" (from the harmonic function theory of Simon Laplace) at 75°E and at 104°W, with two unstable "tesseral points" at 165°E, and at 14°W. Regular fuel consuming corrections must be done from the longitude station keeping.

2.3 North-South Drift (Orbit Not in the Equatorial Terrestrial Plan)

The launch may be such that the satellite orbit makes a small angle γ with the equator plane. In the figure the intersection of the orbit with the equator is LL' (we assumed for simplicity that the apogee of the orbit was at the nominal longitude). The maximum north latitude deviation is ?at 0H and south at 12 h. It remains north from A (apogee) to L' and south from L' to P (perigee). The time during which there is a north-south drift is also computed numerically in Table 1, for the north drift it is the time when $\theta = \pi/2$ starting from the apogee P at 0 h. For $e = -0.507$ the satellite has a 0° latitude at 9 h 42 m and 14 h 18 m.

2.3.1 North-South Deviation during the Day

The north-south drift is the "declination" δ of the satellite (corresponding to VM) for a given angle θ. The spherical triangle $L'VM$ (with a rectangular angle V such as $\sin V = 1$) has the well-known spherical trigonometric relation:

$$\sin \gamma / \sin \delta = \sin V / \sin \left(\frac{\pi}{2} - \theta \right) \tag{8}$$

$$\sin \delta = \sin \gamma \cos \theta$$

("declination" δ of satellite as a function of its angular position θ).

2.3.2 Long-Term North-South Drift

The combined effect of the sun gravitation mainly, of the moon and (the earth potential not being spherical being the least important) tends to "pull" the orbit plane to align with the ecliptic plane earth-moon-sun at a speed of about 1°/year. The fuel budget to maintain the latitude is much more than the longitude station keeping (about 50 m/sec against 2 m/sec) speed impulses.

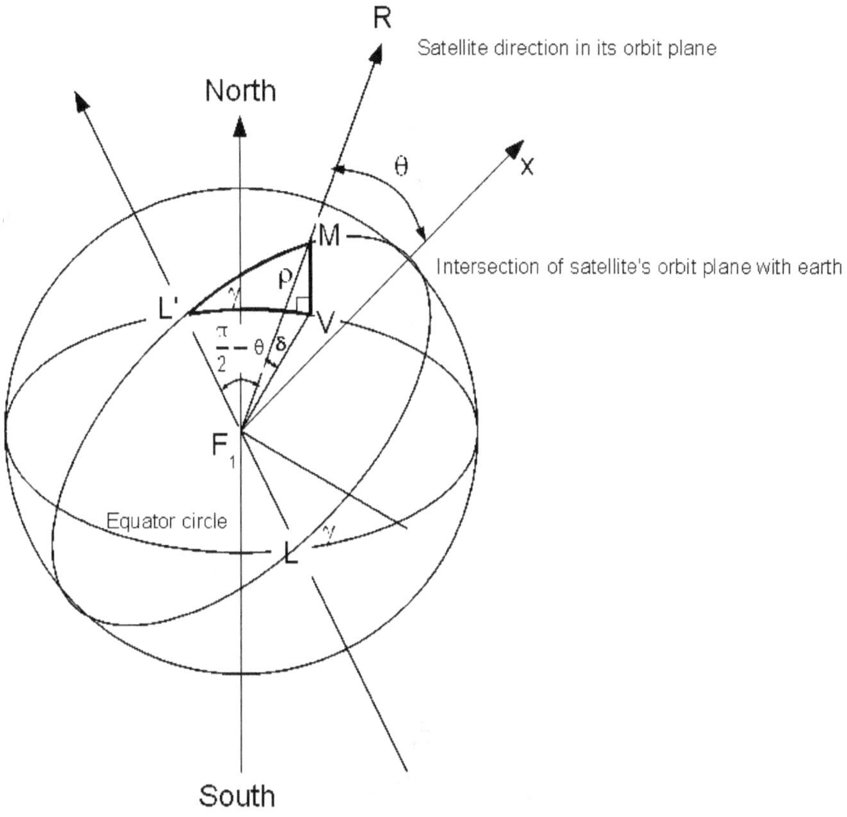

Figure 2 Spherical triangles used in the orbit computation.

2.4 Daily Ground Trace of the "Geostationary" Satellite (Ephemeris)

Not counting the long-term drifts, the ground trace is a combination of the 24 h periods longitude and latitude drifts. In the simplified example (we assume the satellite being exactly at his nominal longitude at the apogee (and perigee). We have, using [7],

- Longitude drift $= (2\pi T/D) - g^{-1}(\frac{A}{DP^2})T$ (difference at time T, between the earth angular position and satellite's which is $\theta(T)$)
- Latitude drift $= \delta(T) = \arcsin(\sin\gamma \cos g^{-1}(T))$

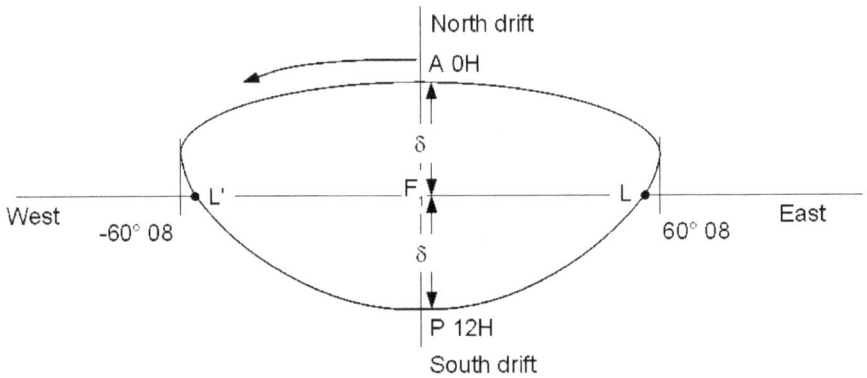

Figure 3 Ground trace of geostationary satellite.

that is, using also [7]

$$\text{Latitude drift} = \delta(T) = \arcsin\left(\sin\gamma\cos g^{-1}\left(\frac{A}{DP^2}T\right)\right)$$

If one wanted to use a satellite to make a traditional computation of its own position with a sextant and accurate watch, the curve gives the ephemeris coordinates α and δ of the satellite.

We gave an elementary but exact 2 body model (earth and satellite) of the orbit computation. In [5, vol. I], there is the more accurate model using the "$n + 1$ body" perturbation theory of Lagrange which is used for long term planet orbit computations and operational satellite ephemeris. When the orbit is close to the simple 2 body model, there is a system of $3n$ second order non linear differential equations, which is numerically computed. (François-Xavier Lagrange who invented the method (1736–1813) must have had a hard time properly solving [he was the first] the 3 body problem earth-sun-moon.)

3 Handover in Satellite Calls

Handover is the procedure which allows a call to proceed without interruption when the satellite handset coverage conditions change due to the serving satellite quick change of position, movement of the handset, or atmospheric conditions.

3.1 Geostationary Systems (Inmarsat (Isatphone), Thuraya)

Each of the satellites may be considered equivalent to a BSC (RNC). Each of the MSC-VLR corresponds to a single ground station with a single BSC. There is a handover when the user moves further East or West than the coverage of the current satellite. The procedure is exactly the standard "inter-MSC handover" [2]. The MSC of the new ground station A sends a MAP_PERFORM_HANDOVER to the previous MSC B giving the next cell where the call must be transferred. It is exactly like GSM and described below for the LEO Inter-gateway handover case.

3.1.1 LEO Systems (Iridium Example)

What is called a "beam" of these systems has a given frequency and covers a "spot" (earth footprint) for the subscribers. Each antenna (e.g. Iridium case) has 16 "beams", so a satellite has a total of 48. Each spot, which may overlap, may be covered by several beams (frequencies). The satellites which have quickly varying positions are connected for control and for relay of communications to a network of XX ground stations, a "satellite gateway", which also are connected to the fixed lines PSTN (calls out and in) and together (inter-MSC handover) by SS7 signalling links through the satellite constellation.

We will explain the LEO handover procedures based on the readers' assumed knowledge of the GSM handover procedures, with a little refresh. For this comparison: "satellite gateway" = MSC-VLR, each satellite = BSC, each "beam" = cell covered by a BTS.

3.1.1.1 Intra-Beam Handover, Equivalent to GSM "Intra-Number Handover". As the quality of the signal degrades, the handset may monitor that another frequency is better ("intra-beam handover" at the initiative of the handset) or the satellite wants he handset to use an other frequency because it may interfere with another satellite's beam. In this case the satellite asks the handset to change the frequency.

3.1.1.2 Inter-Beam Handover [1], Equivalent to GSM "Inter-Number Handover" [2]. The handset decides to switch to another beam of the same satellite because the signal quality is better. In GSM this is equivalent to a handset going to another cell. It sends a request to the BSC and the BSC manages the change including to a cell it has selected.

```
MS            BTS A(beam a)        BSC(satellite)        BTS B (beam b)            MS
before                                                                              after
RR Measurement report
---------------->
                  BSSMAP Measurement Report
                  ---------------------------->
                              Decides to change the BTS
                                    BSSMAP Channel activation
                                    -------------------------->
                                    BSSMAP Channel activation Cnf
                                    <-------------------------
                              The MS changes to the new cell
                              in new BTS
                                                                        RR Handover Access
                                                                        <-----------------------
                                    BSSMAP Handover detection
                                    <-----------------------
                                                                        RR Handover Complete
                                    BSSMAP Handover complete
                                    <-----------------------
            BSSMAP Channel Release
            <------------------------------
            BSSMAP Channel Release Cnf
            ------------------------------>
```

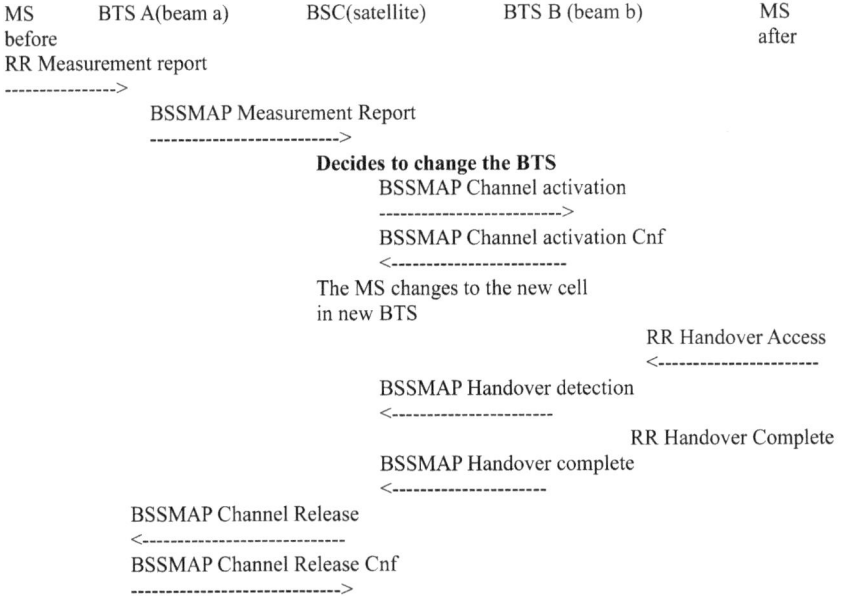

3.1.1.3 Inter-Satellite Handover (Term in [1]), Equivalent to GSM "Intra-MSC Handover" (term in [2]).

The serving satellite is moving and the spot where the handset is making a call may not be covered well any more. The serving gateway A which is used for the call asks a new satellite under its coverage to pursue the call with the handset. The connection of the handset with the previous satellite is released. The gateway acts as a MSC-VLR in "the inter-MSC handover" case (same MSC, new BSC) which is recalled below (it is not standard BSSAP which is used for satellites of course).

```
MS        BSC A (satellite A)     MSC-VLR (ground gateway)      BSC B (satellite B)        MS
before                                                                                     after

          BSSMAP Handover required
          ------------------------------------->
                                    BSSMAP Handover request
                                    ------------------------------------->
                                    BSSMAP Handover request Cnf
                                    <-----------------------------------
          BSSMAP Handover command
          <-------------------------------------------
RR Handover command
<-------------
                                              The MS switches to a new cell and frequency
                                                                 RR Handover access
                                                                 <-----------------------

                                    BSSMAP Handover detection
                                    <-----------------------------------
                                    The MSC-VLR switches the voice circuit
                                                                 RR Handover complete
                                                                 <----------------------
                                    BSSMAP Handover complete
                                    <-----------------------------------
          BSSMAP clear command
          <-------------------------------------------
          BSSMAP clear complete
          ------------------------------------------->
```

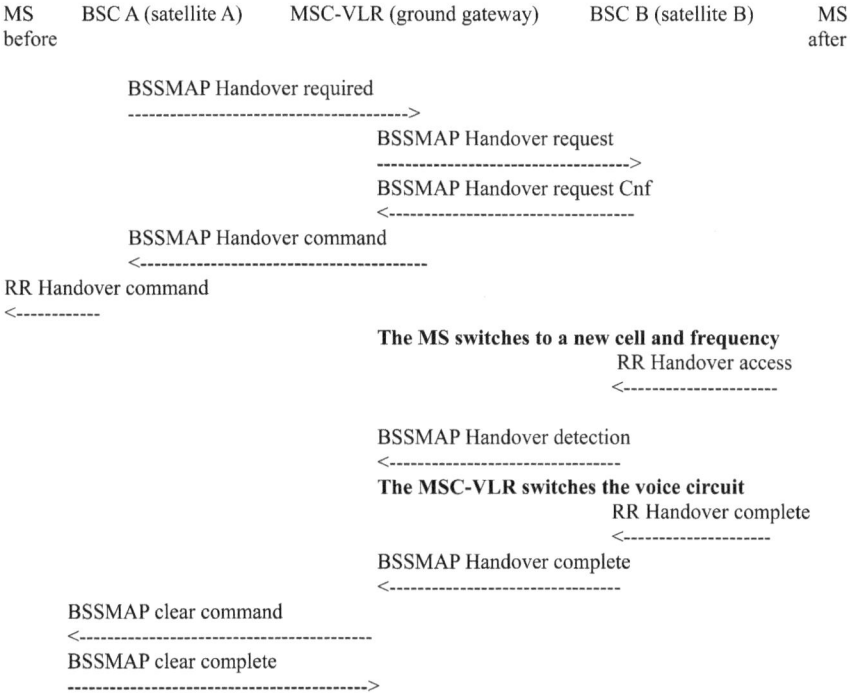

The radio equivalent of the well known SCCP is called the protocol RR. As you can see, there is no use of the MAP protocol as only one MSC-VLR is involved and MAP is used only for Handover between MSC-VLR as in the inter-MSC handover below.

3.1.1.4 Inter-Gateway Handover, Equivalent to GSM "Inter-MSC Handover [2]. Due to the movement of the handset, if it is not well covered any more by a satellite covered by the current ground gateway ground gateway where the call was established ("anchor gateway"). This case exists for satellite but his not described in [1]. When a handset finds a better coverage with an other satellite it will start the same as above, giving the identity of the new satellite (new LAC, Cell Id in GSM) to his current service gateway A.

In GSM it will be as given below:

MS before	BSC A(satellite A)	MSC A	MSC B	BSC B (satellite B)	MS after

BSC A(satellite A) → MSC A:
BSSMAP Handover required
-------------------------------->

MSC A → MSC B:
MAP Perform Handover
-------------------->

MSC B → BSC B (satellite B):
BSSMAP Handover request
----------------->

BSC B (satellite B) → MSC B:
BSSMAP Handover request Cnf
<--------------

MSC B → MSC A:
MAP Perform Handover Cnf
<--------------------

**MSC A establish
a voice circuit to MSC B**

MSC A → BSC A(satellite A):
BSSMAP Handover command
<------------------------------

BSC A(satellite A) → MS before:
RR Handover command
<------------

**The MS switches to a new cell
and frequency**

BSC B (satellite B) → MS after:
RR Handover access
<----------------------

BSC B (satellite B) → MSC B:
BSSMAP Handover detection
<------------------

MSC B → MSC A:
MAP Process_Access_Signalling
<--------------------

MS after → BSC B (satellite B):
RR Handover complete
<----------------------

BSC B (satellite B) → MSC B:
BSSMAP Handover complete
<------------------

MSC B → MSC A:
MAP Send_End_Signal
<--------------------

**MSC B confirms the voice circuit
creation to MSC A**

MSC A → BSC A(satellite A):
BSSMAP clear command
<----------------------------

BSC A(satellite A) → MSC A:
BSSMAP clear complete
---------------------------->

**The voice calls continues with
A being the "anchor gateway" and
B the "serving gateway"**

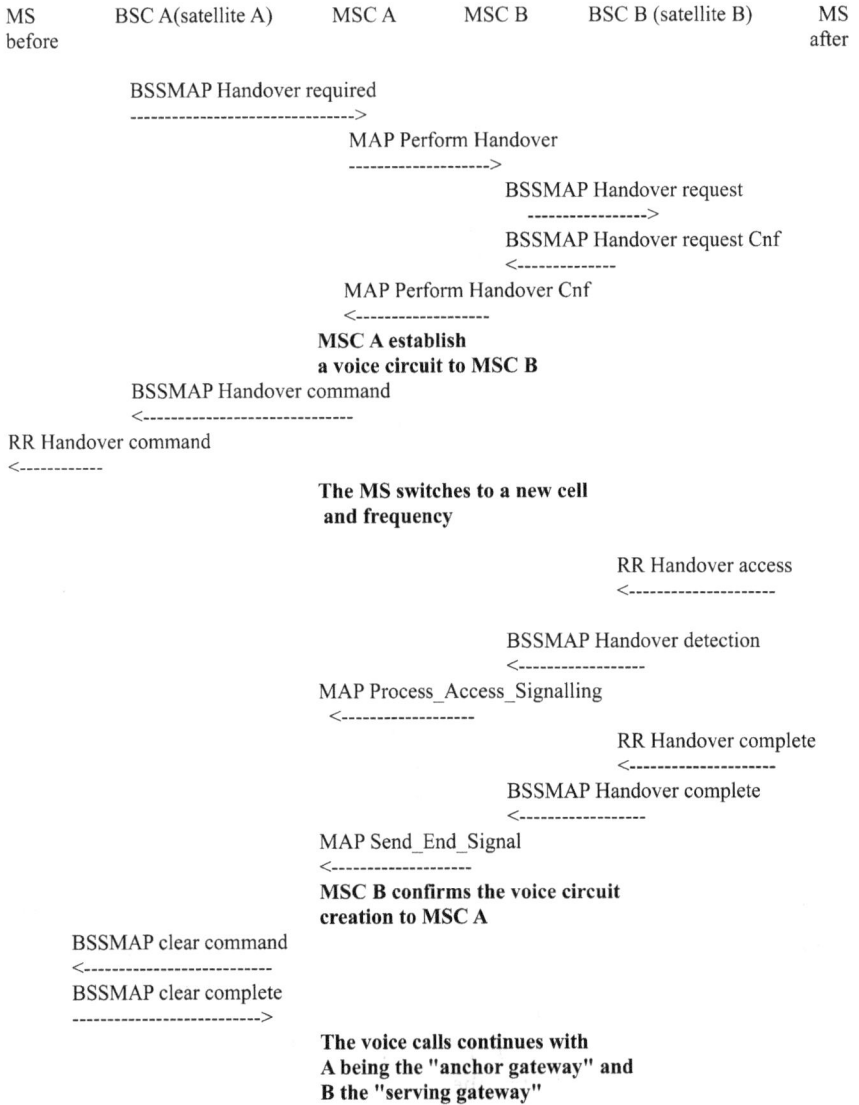

The call is "tromboned" using the voice pass through the satellites as illustrated in Figure 4.

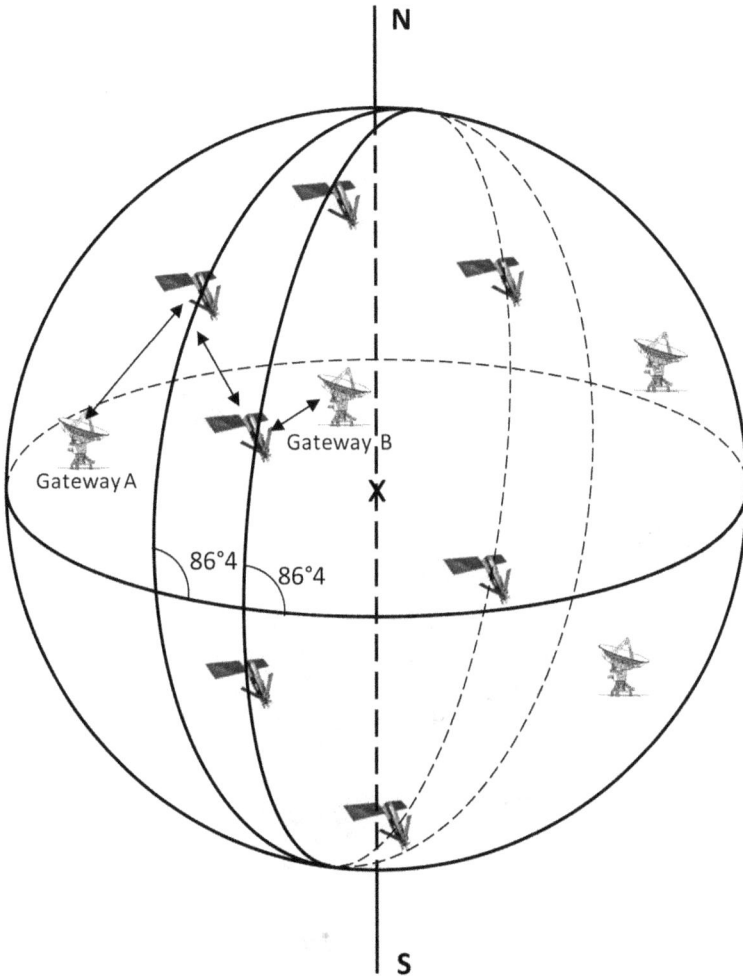

Figure 4 Example of a satellite constellation.

4 Satellite Operators Concerned by the Number Continuity Service

They provide special handsets with a direct radio connection to satellites. In most cases they also operate a satellite network, either geostationary (GEO) with an equatorial orbit or a constellation of Low Earth Orbit satellites with inclined orbits. Below the explain the basis of satellite dynamics and also the principle of telecom space networks with a constellation of LEO spacecrafts when the handover is required to maintain the connections (Iridium, Globalstar).

4.1 Some Satellite Handsets and Core Networks

They are easily recognizable by their voluminous foldable antenna. To be practical, it is easy to see that the handsets below have big antennas, they do not look as compact GSM.

GLOBALSTAR Qualcomm
GSP-1700 CDMA
no SIM card,
no CDMA radio compatibility

GLOBALSTAR Telit
SAT550 GSM (with SIM card)
no GSM radio compatibility

INMARSAT Isatphone
Pro GSM with SIM card
no GSM radio compatibility

IRIDIUM 9555
Satellite GSM (with SIM card)
no GSM radio compatibility

THURAYA Hughes
7101 GSM (with SIM card)
has a GSM radio interface

The SIM card terminals accept GSM cards from networks if there is roaming agreement. The user is invoiced by his own network. Only the Thuraya handset is usable on fixed GSM network,

- with GSM or CDMA compatibility (roaming possibility with terrestrial networks):
  - Globalstar (LEO constellation) (GSM with SIM card or CDMA, only the CDMA handset is commercialized any more) (commercialization and national gateways, e.g. TESAM (GSM) France closed in 2001, see list in above tables).
  - Thuraya (GSM).

Table 2 List of satellite operators.

Name	Type of orbits	Number of satellites	Core network technology and vendor	Ground gateways (with GMSC)	Services provided
Globalstar	LEO	48	GSM (Alcatel) and IS-41(CDMA) DSC Communication Corporation	France (Aussaguel), Russia, Chili, Turkey, USA (Texas)	Voice SMS Data
Iridium	LEO	66	GSM (Ericsson)	Tempe, Arizona Wahiawa, Hawaii – owned by DISA Avezzano, Italy, Pune (India), Beijing (People's Republic of China), Moscow (Russia), Nagano (Japan), Seoul (South Korea), Taipei (Taiwan), Jeddah (Saudi Arabia), Rio de Janeiro (Brazil)	Voice S Data
Thuraya	GEO	2	GSM (Hughes)	Emirates	Voice SMS Data
Inmarsat (Isatphone)	GEO	4	GSM (Ericsson 3G HLR-MSC-SGSN-GGSN+ Lockheed Martin)	Bochum Hawai + 1	Voice S Data
AcES (closed)	GEO	1	GSM	Indonesia, India, Taiwan	Closed

- Innmarsat (C and M terminals), and Isatphone handsets with SIM card.
- Thuraya (GSM) with SIM cards.
- AcES (GSM) with SIM cards (closed in 2004), service taken by Inmarsat with new satellites and ground infrastructure under the Isatphone brand.

The GSM handsets have a SIM card and are dual mode GSM and satellite radio transmission. Technically (Thuraya does it), they can visit a terrestrial GSM network if they have the roaming agreement.

- without GSM or CDMA compatibility (cannot use terrestrial networks)
  - Iridium (constellation)

4.2 Operators Not Concerned by the Need for Number Continuity: Air, Sea (Maritime) and GSM "Bubble Service" and Satellite Operators without Voice Services

Contrary, ONAIR, AeroMobile, Meagafon which provide the GSM service (with "femtocells" in the aircraft or ships are not concerned by "number continuity" unless they use WiFi. The ships are equipped with classical BTS and leaking lines are used as antennas. There is a satellite radio link (Inmarsat mostly) with the ground segment. This allows using the standard GSM handsets. Astrium is a "GSM bubble service" provider; they install ground BTS with satellite links (their Astrium or Inmarsat) to their core GSM network (HLR, MSC and IN) .

On the GSMa site (2012), in the categories Air there is: ONAIR, Aeromobile and MegaFon. In the Sea category: 7 operators, Maritime Partners, ATT (Wireless Maritime), Siminn (On Wave), ONAIR(OnMaritime), Seanet, Smart Coms(Blue Ocean), Telecom Italia. Also even if Intelsat is a major satellite operator, they are not concerned (no handset) neither the access providers such as Satcom.

However the Air, Sea and "bubble" operators are concerned by the alternative use of WiFi by smartphones to provide the same service as GSM.

Also even if Intelsat is a major satellite operator, they are not concerned (no handset) neither the access providers such as Satcom.

Table 3 Satellite vendors for the satellite telephone service.

Networks	Satellites	Designer/vendor of mobile
Globalstar (roaming entre les Gateways Globalstar): on peut avoir un numéro +336400x et s'en servir en Australie	52 Loral satellites (originally) 32 Thalès Alenia (new generation), inclined 52° in 8 orbit planes	Qualcomm (based on a CDMA handset, no Globalstar SIM card, the MSISDN Globalstar is wriiten in the handset Telit (GSM)
Iridium	LEO (near polar, inclined 86°4), only system covering polar regions; 66 satellites in 11 orbit planes Thalès Alenia (new generation)	
Thuraya	Boeing (2 satellites)	ASCOM and Hughes
Inmarsat	3 Astrium EADS (4 I-4 and 7 I-2 or I-3)	Elcoteq (Estonia)
AcES (closed 2011) had roaming with: Hong Kong CSL Hutchison Telecom (HK) Bharti Hexacom Ltd (AIRTEL) Excelcomindo (Indonesia) PT Indonesian (INDOSAT) PT Telekomunikasi Selular (TELKOMSEL) Safaricom (Kenya) DiGi Telecommunications (Philippines) SingTel Mobile Singapore Dialog Axiata (Sri Lanka) Swisscom (Switzerland)	"Garuda" was the satellite name	Ericsson

4.2.1 Satellite Data Services (No Voice)

Operator	Type of satellites	Type of service
Intelsat	66 GEO	
Orbcomm	29 LEO (775km). Small satellites (50–120 kg) launched mid 1990s	Small amount of data (messages) and Automatic Identification System (simple devices). All boats in the Vende Globe race are equipped with an IAS which helps to avoid collisions

4.3 Coverage and Details of the Various Satellite Operators

4.3.1 Globalstar

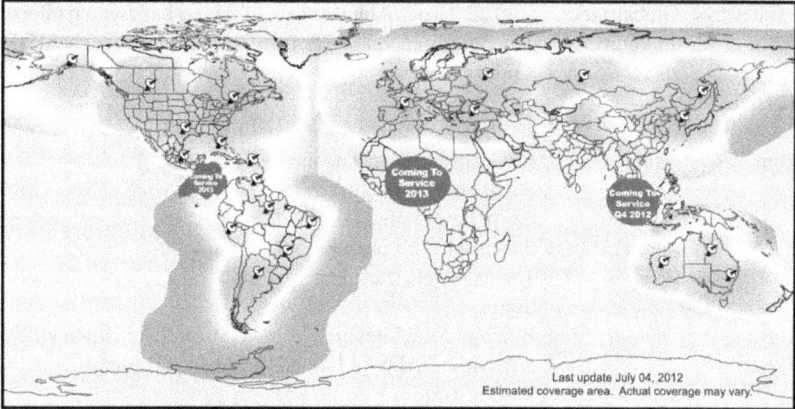

http://www.globalsatellitecommunications.com/globalstar/coverage_map.html

4.3.2 Iridium

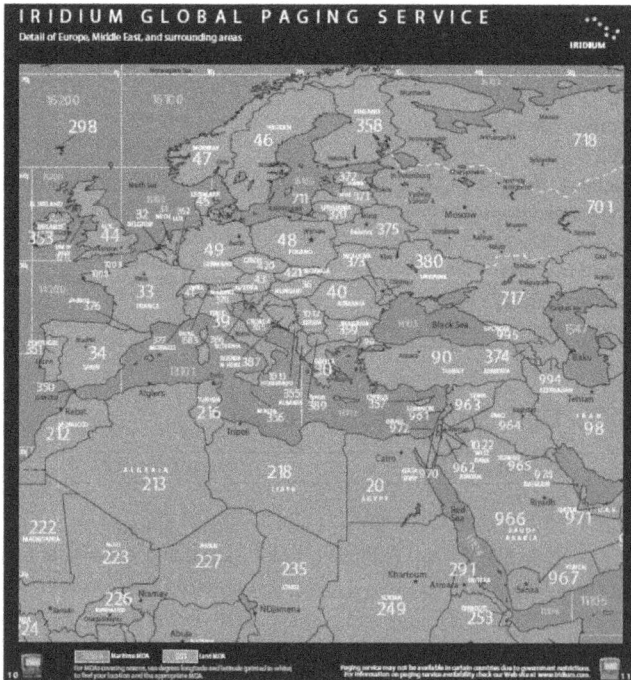

http://www.iridium.com/support/library/CoverageMaps.aspx

- Iridium SSC, Iridium communications service was launched on November 1, 1998. Motorola provided the technology and major financial backing.
- Chapter 11 bankruptcy nine months later, on August 13, 1999.
- Service was restarted in 2001 by the newly founded Iridium Satellite LLC, which was owned by a group of private investors.

The Iridium constellation is the largest in the world, with 66 low earth orbiting (LEO) satellites operating as a fully meshed network. Iridium flexible billing and flat rates for calls from anywhere to anywhere on earth.

GSM-Like Offer
Removable Subscriber Identity Modules (SIMs) are used in Iridium phones, much like those used for GSM. Prepaid SIM cards are usually green while post-paid cards are red.

Iridium operates at only 2.2 to 3.8 kbit/s, which requires very aggressive voice compression and decompression algorithms. Latency for data connections is around 1800 ms round-trip, using small packets.

There is a Web/e-mail to an *SMS gateway* which enables messages to be sent from the Internet or an e-mail account to Iridium handsets for free. There is also a voice mail service.

Tracking Transceiver Units
Without an extra GNSS receiver tracking is difficult, but not impossible, as the position of a mobile unit can be determined using a Doppler shift calculation from the satellite. These readings however can be inaccurate with errors in the tens of kilometres. Even without using Doppler shifts, a rough indication of a unit's position can be found by checking the location of the spot-beam being used and the mobile unit's timing advance.

The position readings can be extracted from some transceiver units and the 9505A handset using the -MSGEO AT command.

4.3.3 Thuraya

Thuraya, is an international mobile satellite services provider, based in Abu Dhabi and Dubai, the United Arab Emirates, and covering mainly the Middle East, Africa, Western Europe, Asia and the Australia.

The system allows telecommunications in voice, data and SMS. The services also provide the GmPRS for direct access to the Internet.

Several models are available; they allow the connection by satellite as well as the GSM networks.

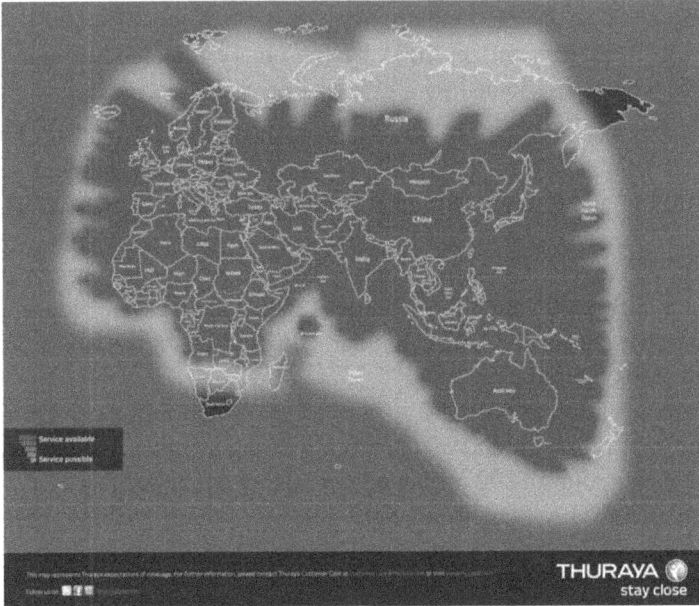

http://www.thuraya.com/coverage-map

4.3.4 Inmarsat

Inmarsat Coverage Foot Print
The MAP gives the number of the "spots" which can be used to get a very rough estimate of the handset position with the timing advance.

Systematic Legal Interception
USA, Russia, India et China enforce that the aircraft using GSM/ Inmarsat, when they fly over their territory have all the voice and data communications "tromboned" through their monitoring system before coming back to the ground station concerned. This is automatic as the satellite links transmit permanently the coordinates of the aircraft and the ground station, which has a numerical map, automatically establishes and suppresses the "tromboning".

The satellites are digital transponders that receive digital signals, reform the pulses, and then retransmit them to ground stations.

Ground stations maintain usage and billing data and function as gateways to the public switched telephone network and the Internet.

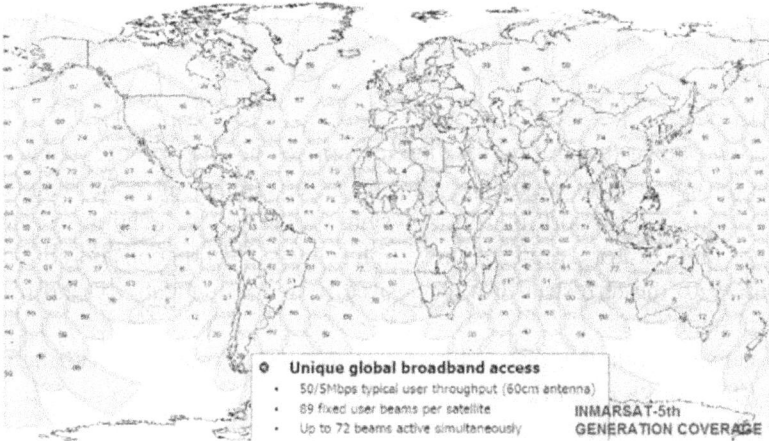

Unique global broadband access
- 50/5Mbps typical user throughput (60cm antenna)
- 89 fixed user beams per satellite
- Up to 72 beams active simultaneously

INMARSAT-5th
GENERATION COVERAGE

http://www.inmarsat.com/cs/groups/inmarsat/documents/document/016329.pdf
http://www.groundcontrol.com/Global_Xpress_Coverage_Map.htm

Teleports and Satellites

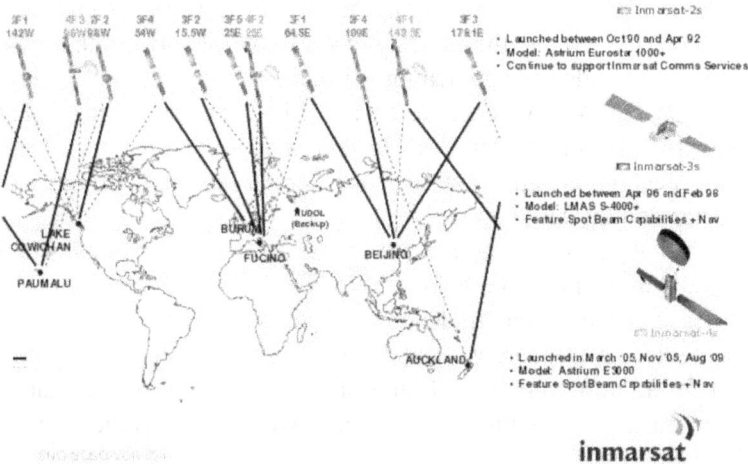

Inmarsat's Fleet and TT&C Network
April 09

Inmarsat-2s
- Launched between Oct 90 and Apr 92
- Model: Astrium Eurostar 1000+
- Continue to support Inmarsat Comms Services

Inmarsat-3s
- Launched between Apr 96 and Feb 98
- Model: LMAS S-4000+
- Feature Spot Beam Capabilities + Nav

Inmarsat-4s
- Launched in March '05, Nov '05, Aug '09
- Model: Astrium E3000
- Feature Spot Beam Capabilities + Nav

inmarsat

Country Codes

The permanent telephone country code for calling Inmarsat destinations is 870 SNAC (Single Network Access Code). The 870 number is an automatic locator; you don't have to know to which satellite the destination Inmarsat terminal is logged in.

Gateways

The fixed part of the ground segment comprises of one or more gateways to access the space segment, transport the necessary signalling, control and communications and support inter/intra system mobility. The location of gateway depends on operator's preferred selection criteria. Factors considered include proximity to the terrestrial traffic, desired network connectivity and routing arrangements.

Gateways comprise of a radio system to support transport over the satellite system and a network switching system to interconnect to the terrestrial network. The satellite mobile phone uses the IP protocol.

The gateways are interfaced to various terrestrial networks such as Public Switched Telephone Network (PSTN), Public Land Mobile Network (PLMN) such as GPRS or UMTS, Internet, private network, etc. The service providers, responsible for end-to-end service provision, access the satellite network through the appropriate interfaces.

BGAN (Global Voice and Broadband Data)

BGAN use the Inmarsat-4 satellites and are used from laptop-sized terminals.

I-4 traffic is transported mainly in the form of IP (Internet Protocol) data packet, thereby increasing the capacity of the Inmarsat network to provide advanced digital mobile communications: Email, Internet access, Secure VPN, Telephony, VoIP, SMS, Videoconferencing, Live video streaming, File transfer, Time critical data transfer, Remote surveillance. The Inmarsat IsatPhone represents a cell phone option at low cost. The service is initially available in Asia, Africa and the Middle East.

FleetBroadband is the tradename of the BGAN technology when it is implemented in the maritime field.

SwiftBroadband is the tradename of the BGAN technology when it is implemented in the aeronautical field (Onair and Aeromobile are using Inmarsat for aircraft communications) Inmarsat Ventures Plc (Inmarsat) did not report having the roaming agreements with other operators except of SMS interconnection.

inmarsat

Land portable

	Wideye Sabre I Voice and data, single-user device	EXPLORER 300 Highly compact, robust device	EXPLORER 500 High bandwidth, highly portable device
Manufacturer	Addvalue Communications www.wideye.com.sg	Thrane & Thrane www.thrane.com	Thrane & Thrane www.thrane.com
Size	259 x 195mm (1.6kgs)	217 x 168mm (1.4kgs)	217 x 218mm (1.4kgs)
Standard IP	Up to 240 / 384kbps (send / receive)	Up to 240 / 384kbps (send / receive)	Up to 448 / 464kbps (send / receive)
Streaming IP (send & receive)	32, 64kbps	32, 64kbps	32, 64, 128kbps
Voice	Via RJ-11 or Bluetooth handset / headset	Via RJ-11 or Bluetooth handset / headset	Via RJ-11 or Bluetooth handset or 3.1kHz audio / fax
ISDN	N/A	N/A	64kbps via USB
Other data interfaces	Ethernet, Bluetooth	Ethernet, Bluetooth	Ethernet, USB, Bluetooth
Ingress protection	IP 54	IP 54	IP 54

Inmarsat and KPN have agreed that all SCCP traffic towards Inmarsat is routed via KPN. For this purpose the INMARSAT global title range for SCCP routing is linked to the KPN network. The KPN and Inmarsat networks are interconnected for the distribution of SCCP traffic.

This document will describe the number ranges involved, the translation rules, the platform releases and contact persons. Note that this document only relates to SCCP traffic for SMS, since Inmarsat has not implemented any roaming due to both terminal and air-interface incompatibility with other terrestrial based GSM/3G operators.

It is not required for a network operator to have an AA.19 interworking agreement with either KPN or Inmarsat for sending SMS traffic towards Inmarsat subscribers.

5 Ground Implementation of the Number Continuity

5.1 Case of Standard GSM Core Network

In [4, chapter 2], one finds the very detailed message flow for a classical single or multi-imsi virtual roaming service implementation. Figure 5 is the signaling flow for Globalstar's GSM based handsets, which use a GSM core network. It would be the same thing for Iridium, Inmarsat or Thuraya which also have a GSM core network. The Roaming Hub has a table to correspond the "IMSI auxiliary" (received from the satphone) to the "IMSI nominal" (the GSM). In (1) , the VLR sends a SEND AUTHENTICATION to request an authentication "challenge". To secure the system, this is *looped back (2) to the Globalstar HLR*, which returns a Challenge (RAND and SRES) back to the satphone (3) and (4) through the Roaming Hub. If the IMSI auxiliary in the satphone compute and match the challenge, the signalling will continue with an UPDATE LOCATION req (5) which is transformed by the Roaming Hub (IMSI auxiliary replaced by IMSI nominal). The GSM HLR will transfer in an INSERT SUBSCRIBER DATA the MSISDN (GSM) (7) which is assigned (8) to the satphone.

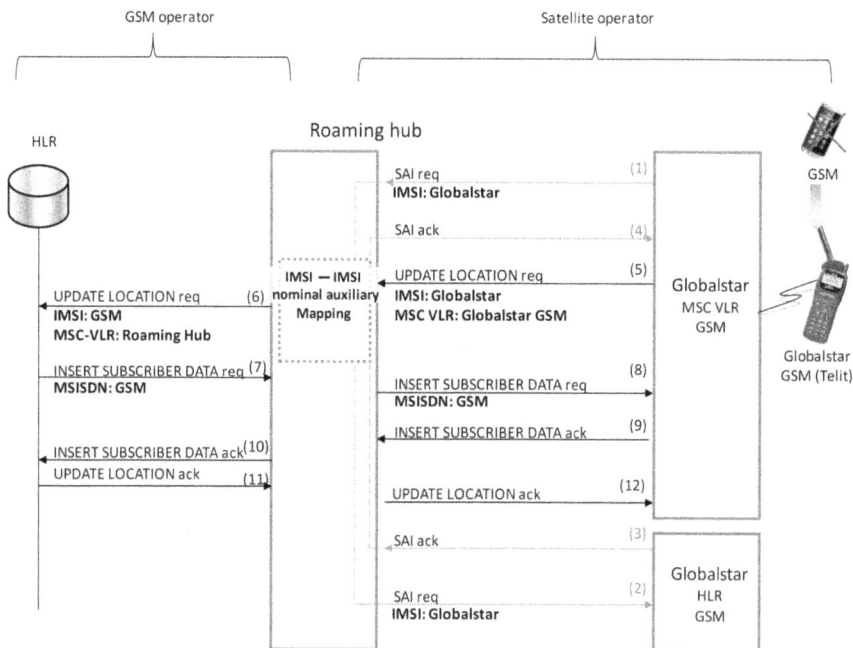

Figure 5 Standard GSM core network.

GSM operator

Satellite operator IS-41

Roaming hub

HLR

AUTHENTICATION REQUEST req   (1)
MIN: Globalstar

AUTHENTICATION REQUEST resp   (4)
RAND, AUTHR

REGISTRATION NOTIFICATION req   (5)

GSM

Globalstar
MSC VLR
IS-41

UPDATE LOCATION req   (6)
IMSI: GSM
MSC-VLR: Roaming Hub
INSERT SUBSCRIBER DATA req (7)

IMSI – MIN
nominal auxiliary
Mapping

MIN: Globalstar
MSC VLR: Globalstar IS-41

Globalstar
CDMA
(Qualcomm)

MSISDN: GSM

REGISTRATION NOTIFICATION resp   (8)
MDN: MSISDN GSM

INSERT SUBSCRIBER DATA ack (9)
UPDATE LOCATION ack   (10)

AUTHENTICATION REQUEST resp   (3)
RAND, AUTHR

AUTHENTICATION REQUEST req   (2)
MIN: Globalstar

Globalstar
HLR
IS-41

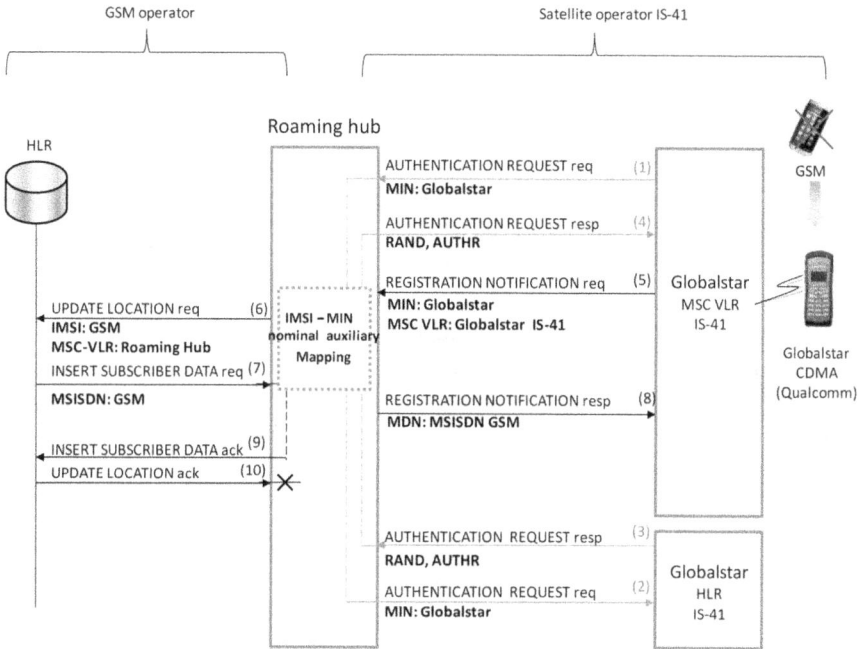

Figure 6 IS-41 core network.

5.2 Case of IS-41 Core Network

It is the same signaling flow as for GSM core networks, except that the Roaming Hub must translate certain IS-41 messages to their GSM counterpart. Figure 6 follows the same pattern and gives a quick equivalence (MIN is the IS-41 equivalent of IMSI). Here we only give the principle of the authentication and registration step, in future work we will show in particular the detailed message flows for voice calls and SMS.

References

[1] P. Nicopolitidis, M. S. Obaidat, G. I. Papadimitriou, and A. S. Pomportsis. Wireless Networks, Chap. 7, Wiley, 2003.
[2] X. Lagrange, P. Godlewski, S. Tabbane. Réseaux GSM-DCS, Chapters 5 and 10 ("Handover"), Hermès, 1997.
[3] Wolfram Mathematica website, integrals.wolfram.com (Note: this gives a real value function only for $|e| > 1$).
[4] A. Henry-Labordère. Virtual Roaming Systems for GSM, GPRS and UMTS, Wiley, 2009.

[5] Maurice Roy. Mcanique, Vol. I Corps Rigides, Vol. II Milieux continus, Dunod, 1965 (in particular, Vol. I, pp. 62–65).

Biography

Arnaud Henry-Labordère is a graduate engineer from Ecole Centrale de Paris (1966), Ph.D. (Mathematics, USA, 1968). He was professor of Operations Research at Ecole Nationale des Ponts et Chaussées during 25 years, as well as at Ecole Nationale des Mines de Paris. He is currently Visiting Professor at Prism-CNRS. He started at IBM research (1967) and founded three companies: FERMA (voice mail systems in 1983), Nilcom (first SMS network in 1999) and currently Halys (telecom equipments). He is the author of eight books (six in maths, two in telecoms) and has been granted 85 patents.

Mobility and Spatio-Temporal Exposure Control

Exposure Control as a Primary Security and Privacy Tool Regarding Mobility, Roaming Privacy and Home Control

Geir M. Køien

University of Agder, Norway; e-mail: geir.koien@uia.no

Received 15 January 2013; Accepted 17 February 2013

Abstract

Modern risk assessment methods cover many issues and encompass both risk analysis and corresponding prevention/mitigation measures. However, there is still room for improvement and one aspect that may benefit from more work is "exposure control". The "exposure" an asset experiences plays an important part in the risks facing the asset. Amongst the aspects that all too regularly get exposed is user identities and user location information, and in a context with mobile subscriber and mobility in the service hosting (VM migration/mobility) the problems associated with lost identity/location privacy becomes urgent. In this paper we look at "exposure control" as a way for analyzing and protecting user identity and user location data.

Keywords: exposure control, vulnerability, risk, identity privacy, location privacy, home control, mobility, cloud, roaming privacy.

1 Introduction

Controlling the degree of "exposure" is one way to reduce risk. If secret and/or sensitive information is exposed then it is more susceptible to being exploited in some way. If we can reduce or eliminate the exposure then the corresponding risk will be reduced or even eliminated (for that particular case).

In this paper we will investigate the concept of spatio-temporal exposure control. Our contexts is users on-the-move (mobile phone/laptop/pad). However, these days it may not only be the user that is on-the-move. Hosted services may also be on-the-move, and cloud services are an example of this. VM migration is already a well-established concept and VM mobilities have also been proposed and discussed in the literature.

Some services may then even move along with the user. In the future one may even subscribe to "follow-me" services, though the most likely seems to be that "follow-me" would be a quality-of-service attribute. Services that need low-latency and services that needs to stay within the same jurisdiction as the user may benefit from a "follow-me" feature.

Mobility is the order of the day, and we should expect this to affect the intruder(s) too. In a geographically distributed environment it may be necessary for the intruder to move alongside its targets, or otherwise it may fail to intercept communications, etc. In this respect it is important for the intruder to be able to distinguish users and services uniquely, and so it will be a goal for the intruder to obtain tracking references to the various objects and entities.

1.1 Exposure Control

Risk analysis methods and the corresponding countermeasures and mitigation is an important part of systems design, configuration and deployment. Modern methods like the TVRA methodology (see Section 2) represent a fairly complete approach to risk assessment, but there is still room for improvements.

The "exposure" an asset experiences plays an important part in the risks facing the asset. The exposure is, technically speaking, not a risk, but it certainly can put vulnerabilities and weaknesses into focus. Thus, increased exposure will increase the probability that vulnerabilities and weaknesses are uncovered. In this context we propose that exposure control mechanisms will be a useful tool in controlling the risk.

1.2 Home Control

The "Home Control" concept originates with cellular operator community and has had a particular standing within the North American operator community [5]. The basic problem that faced the cellular operators was that the classical roaming model was, with regard to trust, a very naive model.

The cellular roaming model is a model with extensive delegation of responsibilities to the visited network. The delegation even extends to the authentication and key agreement protocol; the sessions security credentials are simply forwarded to the visited network. Even worse, the forwarding of the security credentials is potentially not bounded to any authentication event. That is, the visited network may receive the credentials at time T1 and only use the credentials at time T2. The home network is normally not alerted to the authentication event at time T2, and thus the home network is functionally offline with respect to authentication of the subscriber [4].

Needless to say, the cellular model leaves a lot to be desired with respect to home control; the home network, and for that matter the subscriber, must trust the visited network to an unreasonable degree with respect to incurred charges from service consumption, etc. The home network has almost no way of verifying that the subscriber has consumed the services since it very seldom is in direct contact with the subscriber when the subscriber is roaming.

This is unsatisfactorily seen from the home network perspective and there is a clear need for the home network to have more control over the authentic-ation. Tighter control over the associated charging is in place, one option is to require near real-time exchange of charging data, but this is still a reactive fact measure.

Home control classification:

- *Pro-active Home Control*
  Deployment of strong 3-way online authentication is a pro-active secur-ity mechanism. Access control functionality is another example. Other schemes that aim at prevent problems from ever occurring would also be classified as a pro-active mechanism.
- *Re-active Home Control*
  Real-time charging and anomaly detection schemes is a re-active secur-ity mechanisms. Basically re-active mechanisms must have a strong and focused detection capability in addition to an ability to react adequately to the detected incident.

Both pro-active and re-active schemes will have their place in a security architecture.

1.3 Home Control for Cloud Services

The home control concept found in the cellular roaming context has also relevance for cloud services. For instance, in a public cloud environment the cloud service operator has a similar role to the visited network in a cellular environment. The subscriber and the home network will be similar to the VM initiating user and the organization he/she is associated with (and which has the agreement with the cloud service operator). The mapping is the following:

- Cellular Subscriber $\cong$ VM initiating user (USR)
- Home Network $\cong$ Service Subscriber Entity (SSE)
- Visited network $\cong$ Cloud service operator (CSO)

The Service Subscriber Entity may be identical to the VM initiating user, but it may for instance also be the employer of the user.

The problems with lacking home control is also quite similar, and we note that in a basic configuration the USR/SSE has very little control over the submitted VM, the associated data and the outcome of the VM execution. To some degree the problem can be solved the same way as one did for the cellular roaming case, namely to more or less blindly trust the cloud service operator to protect the program/data and to carry out the requested operations as intended. However, it should be clear that while the naive trust model underlying cellular roaming have worked well it is quite inadequate for many, if not most, scenarios which involves processing of confidential and/or otherwise sensitive data.

1.4 Spatio-Temporal Contexts and Mobility Model

The overall context described and discussed in this for mobile/cellular subscriber and for mobile hosted services. The mobile hosted services are VM based services where one may expect VM migration or even VM mobility. The intruder may also be mobile, or even geographically distributed. Figure 1 depicts a possible mobility model. With respect to temporal issues we expect all contexts to be temporally contained, but also that context renewal is possible.

1.5 User Privacy and Identity Protection

Thus, for both cases we have that the location is variable parameter. Another part of the context for our investigation is user privacy. Given that we deal with mobility it is no surprise that location privacy is of interest, and associ-

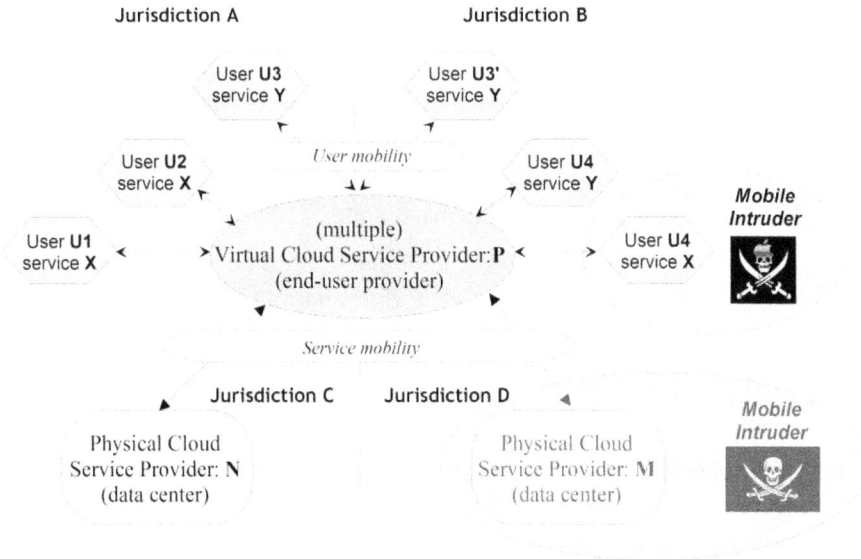

Figure 1 Generic 3-way mobility model.

ated with it we have identity privacy. Data privacy, transaction privacy, etc., also comes into play, but in this paper we primarily deal with identity and location privacy. Privacy, of course, may very well be an end to itself, but we note that lack of credible privacy may easily lead to other security problems.

Identity theft is a growing concern and while it may have received more attention a few years ago, it really does have an impact. According to *The New York Times* (2011/02/09, "The Rising Cost of Identity Theft for Consumers" [19]) the reported number of incidents in the U.S. actually fell in 2010 by approx. 28%, but the cost associated with identity theft still rose. A staggering 8.1 million adults in the U.S. were victims to identity theft and the associated cost has been estimated to be approximately $631 on average. This number did rise sharply from 2009 when the cost was only $387 on average, and the total is now in excess of $5 billion. Similar numbers have been reported elsewhere and in the U.K. the reported numbers were an accumulated cost in excess of £2.7 billion and it affected more than 1.8 million people [20].

Measures that reduce the risk of identity theft therefore clearly seem worthwhile and to limit the exposure seems indeed to be a useful approach.

1.6 Identity Theft

Identity theft is not specifically considered in this paper. Suffice it to say that if one looks at the impersonation aspects of identity theft, it should be clear that a successful identity theft scam relies upon two factors:

- Knowing the identity/identifier of an entity,
- Convincingly claiming to be the said entity.

To prevent impersonation/masquerade one must then prevent the intruder from learning the identity/identifier and/or preventing the intruder from being able to corroborate the identity/identifier.

From a security perspective alone it is not important to conceal the identifiers, in fact identifiers are almost always presented in plain text in authentication protocols presented in the literature [33]. Strong authentication will effectively prevent masquerade. By strong authentication we must here require that a security context is set up by the authentication procedure and that key material associated with the context is used thereafter to cryptographically protect all transactions between the parties.

From a privacy point of view one should obviously not leak privacy sensitive information like an identity. Short-lived transient identities may not matter that much, but permanent or long-lived identifers may allow an intruder to track the target entity. Of course, to claim that exposure of short-lived identifiers does not matter requires qualification. What is short-lived supposed to mean? Furthermore, we *must* require that there is no apparent correlation between the various identifiers used by the same entity. To have a string of short-lived but obviously connected identifiers will not do, as the emergent property would be that of a long-lived identifier.

We should also mention that one may benefit security-wise too from not exposing the identifiers unduely. This is mostly due to imperfect security mechanism, implementation weaknesses and system architecture constraints that sometimes will allow an intruder to potentially gain a weak advantage if he/she knows a subscriber identity. We therefore claim that identity exposure control will also have tangible security benefits as a defense-in-depth type of protection scheme.

2 Brief Introduction to Threat Vulnerability and Risk Analysis

2.1 Next Generation Network

In order to put "exposure control" in context we shall briefly investigate the Threat Vulnerability and Risk Analysis (TVRA) [1] concept. The TVRA methodology was developed by the ETSI TISPAN project for the so-called Next Generation Network (NGN) architecture. ITU-T defines NGN to be:

> A Next Generation Network (NGN) is a packet-based network able to provide services including Telecommunication Services and able to make use of multiple broadband, QoS-enabled transport technologies and in which service-related functions are independent from underlying transport-related technologies. It offers unrestricted access by users to different service providers. It supports generalized mobility which will allow consistent and ubiquitous provision of services to users.
>
> `www.itu.int/ITU-T/studygroups/com13/ngn2004/`
> `working_definition.html`

The ITU-T defined NGN systems architecture has and will have a huge influence on the major core networks and the main access networks.

2.2 Critical Infrastructure Protection

The NGN concept must also be seen in a societal context with increasing dependency on information and communications technology (ICT). In this context the resilience and dependability of the NGN infrastructure becomes crucial, so much so that one has defined the concept "critical infrastructure (CI)". A number of papers and reports has been written about critical infrastructure protection (CIP) and Elsevier has even launched a scientific journal catering to this topic (*International Journal of Critical Infrastructure Protection* (IJCIP)). This paper is not about CIP per se, but we note that exposure control can easily fit into the overall CIP concept.

2.3 Vulnerabilities, Threats, Risks and Threat Agents

Given that an NGN infrastructure is a critical component in a modern society it is necessary to ensure that it is dependable and secure. Within the ETSI TISPAN project one has developed a new methodology for analyzing

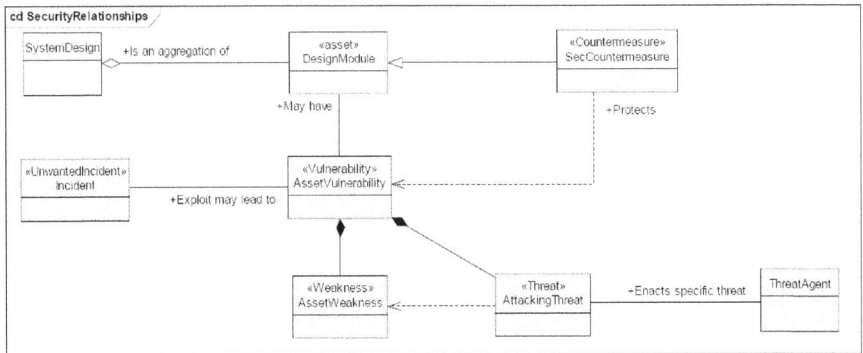

Figure 2 Generic TVRA model.

vulnerabilities, threats and risks associated with NGN type of networks. The TISPAN initiative also encompasses countermeasures and cost-benefit analysis, etc., for the various cases [2]. We shall now briefly outline the TVRA concept [1].

2.3.1 Generic TVRA Model

We have that an *asset* is an object of value that needs to be protected. In a system there may be unwanted/undesireable events concerning an asset. These events are denoted as *incidents*. An incident occurs when a `vulnerability` is exploited. A vulnerability is then a *weakness* which may be *attacked*.

The weakness/vulnerability may exist without there being any incidents, but given a knowledgable *threat agent* the weakness/vilnerability may be exploit and used in an attack. Figure 2, transposed from [1, fig. 4], depicts the Generic TVRA model.

2.3.2 Security Objectives and Threats

In the TVRA model on defines four primary security threats and five primary security objectives. Primary threats:

- Interception,
- Manipulation,
- Denial of Service (DoS),
- Repudiation (sending and/or receiving).

The security objectives do not correspond directly with the threats, but there are obvious relationships. The primary security objectives (known as CIAAA):

- Confidentiality,
- Integrity,
- Availability,
- Authentication,
- Accountability.

2.3.3 Weaknesses, Vulnerabilities and Threats

A precondition for a threat, in the above model, is to have a threat agent. For a large system it is naive not to assume that the threat agent, also known as intruder, adversary, enemy or even hacker, is present. There exist many different types of threat agents, ranging from spectacularly powerful intruders to opportunistic and less resourceful legitimate users that simply try to elevate their access rights beyond what has been agreed. One examples of these is the classical Dolev–Yao Intruder [12] and in [13] one defines a set of intruders based on their capabilities and financial strength. In [14] one further discusses the computational strength of attackers in context with Moore's "law".

So, we assume threat agents to be present. Some of these agents will be powerful and some will be less so, but the lesser agents may be numerous and may in the end prove to be a larger problem for the overall system.

A real-world system will have weakness and vulnerabilities. Some of these weaknesses and vulnerabilities are simply due to weak design or erroneous implementation, while other arise due to inescapable complexities or due to design decision that give priority to certain features over other features. According to Anderson's classical "Why Cryptosystems Fail" [15] one should also assume that quite a few of these weaknesses are due to misunderstood security objectives, to inadequate threat models and to misguided trust assumptions.

Whatever the reason or cause, the vulnerabilities and weaknesses exist in the system and they will be susceptible to exploitation by an threat agent provided that the vulnerabilities/weaknesses are visible to the threat agent. In this context we argue the case that "exposure" should be included as a class of vulnerability and that "exposure control" should be an independent counter-measurement in an extended TVRA method.

2.3.4 Conflicting Incentives

Why do we carry out risk analysis activities? Obviously, to identify risk and to reduce and mitigate it as we see fit. However, what is a risk or liability to one party is an opportunity to another party. In terms of privacy, it should be clear that private information has value to more than one party. Unfortunately, the there will often be clear conflict of interest, and this is dependent for privacy sensitive information.

An example would be web surfing and searching. You may want to remain anonymous while Google, Facebook, Instagram and other services will potentially stand to make profit from knowledge about you and your habits. So there may very well be conflicting incentives during the web transactions. The Conflicting Incentives Risk Analysis (CIRA) methodology is one way to capture this [34].

A risk analysis methodology should be able to capture and cater for conflicting incentives and interests to be truly useful. The TVRA methodology does not currently cover this, but it should be possible to extend it to cater for those needs too, perhaps by including CIRA methods.

3 Mobility and Migration

It goes without saying that cellular subscribers can experience full mobility. Seamless mobility is also a standard service in cellular systems. Traditionally the functionality has been limited to mobile phone handsets, but nowadays mobile termination (MT) units are commonly integrated into laptops, tablets and other gadget. The mobile device may of course also be an embedded device, and we may therefore potentially include all Internet-of-Thing (IoT) devices. The distinction between traditional cellular services and other wireless service are also blurred and more or less meaningless to the customers. Thus, we can safely postulate that users with laptop/smartphone/ipad and other gadgets will, as the default assumption, be mobile subscribers in the sense that they can obtain IP connectivity and that they routinely are on-the-move while being connected.

With the inclusion of IoT devices in the equation we must cover several communications scenarios:

- *Human-to-Human* We note that while the communications may logically be human-to-human it may certainly be conducted and facilitated by mobile devices at the lower layers.
- *Machine-to-Machine* Embedded devices are quite often wirelessly connected. It those cases one must assume mobility to be the norm. As stationary wireless device can safely be modelled as a mobile device with a special case of zero velocity.
- *Human-to-Machine*

The interface must be adapted to humans, but apart from that human-to-device communications need not be special at all. Again, we shall assume wireless communications to be the model.

Technically speaking there isn't a big difference between the mobility handling in the above scenarios. There may be humans involved on causing the mobility, but the technical realization of mobility handling at the lower layers will invariably be handled by some mobility management machinery.

In the reminder of this section we investigate service mobility in the guise as VM migration/mobility.

3.1 Physical VM Migration

Live migration is by now a standard option in most cloud services. Basically it allows a server administrator to move a running VM or application between different physical machines while providing uninterrupted service. Live migration requires that allocated memory, storage, and network connectivity of the VM is successfully migrated to the destination machine. Seamless migration is defined to be a migration event that is transparent to the services consumer.

Migration is normally considered to be a "local" event in the sense that one normally assumes that both source and target machine is physically close to each other, i.e. within the same data center. That is, one can safely assume that normal VM migration is restricted geographically and generally within the same physical premises. So one does not need to worry about switching country or switching host operator, etc.

3.2 Physical VM Mobility and VM Roaming

VM mobility is somewhat more of a novelty, but it is not a new concept [23]. In VM mobility the VM is moved beyond the traditional "local" boundaries and the mobility is not per se limited in physical distance. In practice one cannot have full service continuity for prolonged relocation procedures, but this would of course depend crucially on required service response times and on the quality of the connection (bandwidth and latency).

The case argued in [23] is for very low latency services and where the executing VM needs to be in the physical vicinity of the user in order to minimize network propagation delays. Whatever the motivation, it should be clear that techniques that allow VM migration would also allow VM mobility. The upshot of VM mobility is that one cannot be entirely sure that the VM stays within the same data center and therefore it may potentially move across borders and potentially onto a different hosting environment/service.

We shall denote VM mobility onto a different host environment as *VM roaming*. This will include VM mobility from host A to host B, where host A and B is the same company, but located in different jurisdictions.

3.3 Physical Mobility of the Server

The service platform may itself be a physically mobile platform. Laptops, mobile phones and other gadgets may be used as a host platform. These platforms, while somewhat computationally restricted, are plentiful and are themselves mobile. Social networks or corporate networks may utilize the user client platform to host simple cloud services. These services may be private cloud services and they may be specific to a service, but ultimately they could be realized as publicly available generic hosting services.

3.4 Technology Mobility

As of today the VM technologies are fairly specific both to physical host platform and to hypervisor/VM manager type. However, it is of course possible to fully emulate one environment within another, and so a type-X VM can be run on a type-Y environment provided that a X-to-Y emulation layer exists. It is therefore, in principle, possible to have *VM roaming* cases where the VM moves onto a different platform from where it originated. The technology has not reached that level of maturity yet, but if there are strong enough incentives then surely new technology will be developed to allow this to happen.

4 Exposure Control

4.1 The Case for Exposure Control

The concept of exposure control is not directly linked to weaknesses or vulnerabilities, but obviously the less exposed a weakness or vulnerability is the less likely it is that it can be converted into an attack.

Thus, as a means of "defense in depth" [16] exposure control is about reducing the exposure of assets to a minimum. Defense in depth has not the best reputation in academic papers, but some recent papers analyzing threats and attacks have found that "defense in depth" and "security by obscurity" does have merit in the sense that broad sweeping attacks can be prevented and/or mitigated by these tactics [17, 18]. The reason is found in the cost associated with attacking large populations, but we should warn that these tactics are less likely to be effective against targeted attacks.

Still, we argue, protection schemes that simply aim at concealing the presence or obfuscating the presentee of an entity may be effective against the opportunistic attacks. Since, according to Florêncio and Herley [17] and Pavlovic [18], there is reason to believe that these attack are the most common ones, it makes a lot of sense to employ defense tactics that limit and control the exposure of assets.

4.2 Cryptographical Exposure Control

Exposure control is not a new concept per se and in cryptography and information security there are well developed notion of exposure with respect to cryptographic keys and to the amount of ciphertext that should be encrypted under the same key.

A secret cryptographic key has only a limited amount of entropy and while one relies on effective cryptographic primitives to mask any correlation between the key and plain text data, there will inevitably be information leakage. This leakage cannot be avoided and one must therefore restrict the use of a key so that it does not get too exposed. Exposure and information leakage also happens at the key distribution phase and during storage. The key distribution and key agreement problem can largely be contained with good cryptographic protocols, but the storage problem is harder to solve. A weakness in the hardware platform, any weakness in the system software, the security software or even the application software may leak information about the key. This type of leakage may be entirely independent of the actual usage pattern for the key.

To address these issues cryptographic systems and protocols typically limit the lifetime of secret keys. An example is the IPsec protocol suite where one can limit the "lifetime" for a security association both in terms of usage (no. of bytes/packets) and in terms of passed time (seconds) [21].

In [22] the case is argued for spatio-temporal exposure control and this paper is an important background paper for our investigations.

4.3 Privacy Assets

The primary privacy assets will be permanent identities, the associated location data and of course "data privacy".

Data privacy, in a communications setting, will normally only cover the protocol payload, but what is considered payload is a matter of perspective. At layer **N** the whole of layer **N** + 1 is payload. So, we will refrain from a strict definition, and rather allow "data privacy" to include "all relevant data". With this in mind it is clear that data privacy protection must be implemented sufficiently low in the stack to be able to protect "all relevant data". This is why it is necessary to deploy data confidentiality protection at the link layer in cellular/wireless systems. However, link layer protection is, by definition, limited to the range of the link. Once inside the core network the norm is to protect (aggregated) traffic at the network layer, but this isn't necessarily sufficient as some services are more sensitive than others. That is also why one may need additional protection at higher layers to cover end-to-end aspects.

One may classify the identifiers according to geographic scope and lifetime. An example is shown in Figure 3. A permanent identity will, by definition, by comparatively long lived. It may not necessarily be a secret per se, but is has the potential to be highly privacy sensitive.

Under many circumstances one does not actually use a primary identity for transactions purposes. One may instead use secondary identifiers (numbers, references, addresses) which may be derived from the primary address. Additionally, there may be "emergent" identifiers that may or may not be recognized as identifiers per se, but that may nevertheless be used for tracking purposes by external entities (including our intruder/adversary). Many of these secondary identifiers will be public, but they may also be private. Furthermore, a secondary identifier may have limited lifetime. This may arise out of the given context or it may be by explicit design.

One may also find that there are identifiers that does indicate a class of objects or entities rather than a specify object/entity. However, prolonged use of a class identifier by any specific object/entity may allow for additional data to be associated with the specific instance and so one may in the end derive a unique identity from the context. This "derived identity" may or may not be recognized by the object/entity that it refers to.

Tertiary (transient) identifiers will also exists. These will be short-lived and/or be temporally and/or spatially contained. A typical example would be the M-TMSI temporary identity used in LTE networks [6]. The M-TMSI is unique within the respective MME area, but needs to be qualified for external use (forming a GUTI identifier).

An identity, even a class identifier or a secondary/tertiary identifier, is obviously an assets in our case. The location of the an identified entity/object may also be viewed as an asset. The more precise the location the more valu-

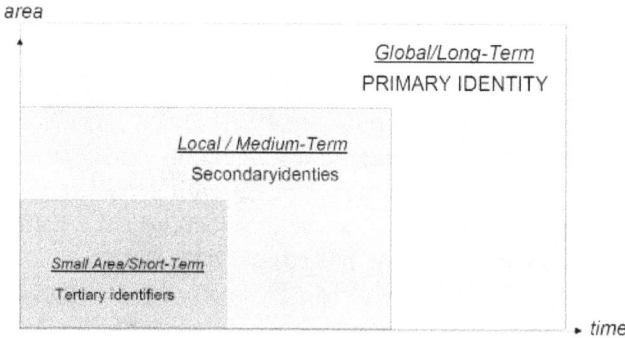

Figure 3 Longevity and scope of identifiers.

able the asset becomes. For mobile subjects to accumulate a time-series of identity/location information is another type of privacy asset, and it is in some cases a more valuable asset. The time-series amounts to tracking information and given tracking information and some traffic data one may easily also arrive at transaction information. Should one be able to also gather user data and associate it with tracking information then one may compose a fairly complete tracking record and this could very well be used for identity theft attacks or similar. As was discussed in Section 1.5 identity theft is a very real threat and the effects of identity theft is amongst the worst both economically and emotionally.

Other identifiers, information and information patterns may also be used as an associated identifier for the subscriber. These may also be abused in identity theft scams. There are many auxiliary identifiers used in a 3GPP system context and they include amongst others the ICCID (smart card ID), the MSISDN number (the phone number) and the IMEI (mobile device serial number), not to mention other non-system identifiers that may be associated with the user/subscriber like various account identifiers (Android/Google, Skype, Facebook, etc.). We shall in this paper limit ourselves to link layer identifiers and then primarily to those associated with setting up an initial security context. The impact of of identity theft, as major source of a privacy intrusion, is discussed in more detail in [24].

5 Privacy-based Spatio-Temporal Exposure Control

5.1 What to Protect

The question is of course not only what to protect, but also from whom one needs to protect the information. When it comes to identity information and to location information we may start off with declaring that "external" non-authorized parties shall not learn neither identity nor location for a user or indeed for an associated (VM-hosted) service. But there are potentially many "internal" parties and not all of them really need to know the privacy assets. So, one must really start off with defining who should have access to the privacy assets. Here we strongly advocate prudence and we advocate that privacy respecting business principles to be used, along the lines of the Privacy-by-Design initiative [25].

5.2 Where Are the Identifers Located

The various identifiers are potentially stored on a lot of different nodes. In a distributed system this means that sensitive information will be stored on nodes in different areas and oftentimes under different jurisdiction. This will make privacy protection complicated, but with respect to enforceability and with respect to trust. To illustrate the complexity we shall briefly outline where some of the 3GPP identifiers are stored. The overview in Figure 4 is by no means intended to be exhaustive. For instance, we have excluded the user equipment (UE) entirely. It must also be mentioned that since the figure contains a mix of 2G (GSM/GPRS), 3G (UMTS) and 4G (LTE/LTE-Advanced) identifiers, credentials and nodes, it is bound to be somewhat imprecise. Inter-generation support (backwards compatibility) for roaming, etc., complicates this picture further.

It should be mentioned too that paging and system access in the 3GPP systems necessarily involves exposure of identifiers. For paging, the identifier will be visible on the over-the-air interface within the whole of the location area (routing area). It should therefore be immediately clear that paging identifiers are widely exposed and that consequently one should *never* page a subscriber with the permanent identity (IMSI). Corollary, there should be a limit to the number of paging events per temporary identifiers too. It should also be clear that quite a few nodes will see privacy sensitive data and that all those nodes must therefore be able to protect the privacy assets.

As Figure 4 shows, the 3GPP systems, with the distributed authentication model, are also vulnerable in that the authentication material and key material

	Serving network					Home
	Cell BTS/NB/eNB	BSC	RNC	MME	VLR/SGSN	HSS (HLR/AuC)
IMSI	Yes	Yes	Yes	Yes	Yes	YES
TMSI	Yes: BTS, NB	Yes	Yes	Not applicable	YES	No
GUTI	Yes: eNB	Not applicable	Not applicable	YES	Not directly applicable	No
MSISDN	Yes	Yes	Yes	Not applicable (except non-ISUP use)	Yes	YES
IMEI	Yes	Yes	Yes	Yes	Yes	Yes
Auth.Info	Challenge-Response data	Challenge-Response data	Challenge-Response data	Full EPS-AV and Legacy contexts	Full triplet and full UMTS AV	YES, ALL
Key material	Yes: BTS,eNB	Yes: GSM	Yes: UMTS	Yes: LTE	Yes: GSM/GPRS and UMTS	YES, but not *AS Security context in LTE*

Figure 4 Distributed identifiers and credentials.

(triplet, UMTS AV, EPS AV) are distributed to the serving network. Exposure of these credentials would make identity theft and impersonation very easy to carry out.

What is not shown in the figure is that the identifers and credential must also be transported between the nodes. So, one also need the communications security to be reliable, available and actually used. In this respect it is clear that the 3GPP security architecture (2G in TS 43.020 [10], 3G in TS 33.102 [10] and 4G in TS 33.401 [10]) is not very strong on requirements on the deployment of the so-called Network Domain Security (NDS/IP) protection (TS 33.201 [9]). There is therefore a high probability that the identifiers and credentials are not protected while in transit. We add here that this may even be the case for information that passes through intermediate networks.

5.3 How to Protect Privacy Assets

Needless to say, this question cannot be fully answered without taking the context into consideration. One needs to define the privacy assets that need protection and one needs to define what it must be protected against and possible also how it is to be used.

For cellular systems one has the potentially conflicting requirement that the home network (HPLMN) needs to have home control while the subscriber (represented by the user equipment UE) needs to have credible privacy. The system access protocols in 3GPP-based system (GSM/GPRS, UMTS, LTE/LTE-Advanced), which includes identity presentation and authentication and key agrement, typically expose both permanent identities (IMSI) and secondary/tertiary identities (TMSI/P-TMSI/GUTI). See 3GPP TS 33.401 [10] for details on the security part of the access procedure.

One also has the serving/visited network (VPLMN), which needs assurance that incurred costs for service provisioning will be accounted for. In [28, 27] this conundrum is discussed and a solution is provided that does indeed provide credible user privacy and a fair amount of home control. Here the basic idea is that mobile device presents itself with an anonymous pseudo-random subscriber identifier, *ASID*, and an encrypted block *A* that contains, amongst others, the long-term identity. Block *A* is encrypted with the public-key belonging to the home operator (HPLMN), and through the AKA procedure the HPLMN will get assurance about its subscriber while the long-term identity is concealed from both the VPLMN and external parties (the intruder). The VPLMN will get confirmation from the HPLMN that the $ASID$ is representing a recognized subscriber and that the HPLMN will accept charging on behalf of the subscriber.

For *location privacy* vs. *home control* one may additionally use Secure Multi-party Computation (SMC) methods to let the home network question the serving network/user about the location while essentially only providing assurance about location without actually revealing the location. In [28] a demonstration of this scheme is demonstrated through a protocol which solves the so-called point-inclusion problem. The protocol in [28] is reasonably efficient (for an SMC protocol), but it is not too practical and it can be circumvented by a dishonest party. Other solutions exists too, and in [3] several of those are discussed.

The cellular system setting can therefore be said to have some solutions and the solutions are even quite good. Other settings which have solutions include IoT-based cases in which a user may access an IoT-based service without revealing too much private information to the IoT device. The actual requirements will dictate the how one solves the problem; Køien [26] provides one example. Another example is found in [31] where one investigates problems associated with privacy and intrusion detection on a mobile broadband platform.

Cloud service privacy is an area where, to the best of the author's knowledge, there is no truly credible and practical solution available yet. The problem is hard in the sense that a VM executing on a remote platform cannot easily verify it own location. Home control for the VM owner is therefore hard to come by. We may attempt to briefly sketch a way forward here, and it seems reasonable to start off with a requirement for *verification*. In cloud parlance this is often described as the *remote attestation problem*. That is, there must be some way for the software-only VM to *verify* its identity (ownership) and its location. The location aspect we may be most interested in here may be the jurisdictional location, but whatever way we choose to classify the location information we still need to have means for verification of (hypervisor) claims. Use of Trusted Computing Module (TCM) functionally [30] may be a way forward, but there nevertheless seems to be a need for at least semi-trust in the cloud service provider. On the subject of trust and cloud services there is actually ways for increasing the trust one may have in cloud services [29]. This does not replace the need for "hard" assurance, but may be a useful addition and a acceptable "defence in depth" addition.

If verification is possible then the next logical step for the VM is to apply that information. Specifically, the VM should now attempt to address and comply with the home control policy. Thus, the VM must somehow be able to *enforce* the home control policy. For instance, if the VM detects that it is executing in a foreign (hostile) legislation it may need to shut down or it may need to set up additional security measures, etc. We believe that if verification is possible then enforcement should be possible too. Another complicating aspect here is that the remote attestation must be conducted for mobility cases. If the VM migrates or is otherwise relocated then clearly one must re-attest the platform. If fact, the re-attestation should be performed before the relocation takes place and it would seem reasonable to assume that the current VM host is responsible for verifying the target VM host before actually moving the VM to the target host. Needless to say, mobility/migration should be subject to policy control and maybe even to some emasure of VM home control.

When it comes to generic identity protection we should of course not forget to mention identity management solutions. Many proposals and initiatives exists and there are also several (national) standards available. One prominent initiative is "The National Strategy for Trusted Identities in Cyberspace (NSTIC)" [32]. For many cases the use of identity management schemes is the only way forward and thus the way privacy and security is handled is of the utmost importance.

5.4 Privacy Policy Control

We believe it is essential to provide some means of privacy-based policy control, which may be used for migrating VMs and roaming subscribers. In the 3GPP framework one already has a general scheme called the "Policy and charging control architecture" for policy control for roaming subscribers (TS 23.203 [7]). The policy control architecture is mostly concerned with charging and QoS aspects. It would be very useful to apply this framework to instruct the roaming partner on how to behave with respect to subscriber privacy.

Exactly how one should best do this remains for further study, but it is immediately clear that policies regarding the frequency of re-assignment of temporary identifiers and of use of NDS/IP protection would be welcome from a security and privacy perspective.

With respect to cloud computing and VM mobility it is clear that it too should be subject to policy control. The VM "handover transfer" should be protected, the target VM should be verified (remote attestation) and the target VM host should be an allowed host. Since tracking of VMs could be an issue it is also important the VM references are anonymity or fully confidentiality protected.

6 Summary

Exposure control is an important aspect of a security architecture. It relates directly to the assets in the system and will as such be part of a generic risk analysis. There will also be exposure control mechanisms, and for privacy these will be associated with various confidentiality services.

Exposure control still needs to be investigated as a part of an extended TVRA methodology. Exposure itself is an aspect of vulnerability and exposure control is a way of mitigating a possible threat. We also advocate to take conflicting interested into account to more accurately reflect the real world. More in-depth work in this direction is recommended.

Exposure control is relative with respect to the asset one is concern about. The asset have different value to the different entities in the system, and there may be conflicting interest here.

In this paper we have primarily looked at knowledge of user identity and user location as the primary assets. Thus, we have in effect investigated privacy exposure control. Privacy is a means to itself and privacy is also a growing concern. Thus, efforts in mitigating privacy problems is clearly is also a means to itself. However, in the literature we saw that privacy intrusions rarely appear to be only an end to itself. In fact, identity theft seems often to be the attacking purpose. That is, identify theft is again the starting points for fraud at large. This is not only a theoretical concern and identity theft related crime has accumulated costs in multi billion dollar region.

For cellular subscriber and for users that needed IoT-based service access we found that solutions existed that would permit credible protection of identity and location. We also saw that identity management solutions exists, and when applied correctly these may indeed also provide a measure of privacy. We note that the user/subscriber does have its own hardware (the mobile phone/device, trusted smartcard (SIM,UICC/USIM)) and that this obviously helps the assurance. Not that one should be too naive here as malware may corrupt the platform, but it obviously is possible to have higher assurance levels when one has control over the hardware platform.

To some extent all the solutions require a level of trust in the participating parties (not always all of them), and this points to the fact the trust management (and associated enforcement mechanisms) is also needed alongside with identity management.

For cases where the mobility is in the hosted service, e.g. in VM provided services, the case is more worrying. Data confidentiality, VM referential identity and VM location may all be privacy sensitive and the VM owner will need a level of "home control" over the VM. Exposure control in this setting is difficult since the VM is all software based and the VM is hosted on hardware which is not under control by the VM or the VM issuer. Thus, there appears that there is no viable way to establish the current status (verification) or to enforce a particular privacy and security policy. That is, use of trusted hardware at the cloud service provider may allow some home control. It should be possible to have remote attestation (verification) and it may be possible to even have a certain level of enforcement.

We believe further research is necessary to conclude on this and we believe that remaining exposure issues can at least partially be solved or mitigated by providing credible trust management solution in conjunction with identity management solutions that emphasizes privacy. Since we are dealing fraud and crime with multi billion dollar interest it seems that contractual matters and legislation that favors safe business conduct also needs to be in place.

References

[1] ETSI, TS 102 165-1. Telecommunications and Internet converged Services and Protocols for Advanced Networking (TISPAN); Methods and protocols; Part 1: Method and proforma for Threat, Risk, Vulnerability Analysis. March 2011.

[2] ETSI, TS 102 165-2. Telecommunications and Internet converged Services and Protocols for Advanced Networking (TISPAN); Methods and protocols; Part 2: Protocol Framework Definition; Security Counter Measures. February 2007.

[3] G. M. Køien. Entity Authentication and Personal Privacy in Future Cellular Systems. River Publisher, Aalborg, Denmark, 2009.

[4] G. M. Køien. An introduction to access security in UMTS. IEEE Wireless Communications, 11(1): 8–18, February 2004.

[5] G. Rose and G. M. Køien. Access security in CDMA2000, including a comparison with UMTS access security. IEEE Wireless Communications, 11(1): 19–25, February 2004.

[6] 3GPP. TS 23.003 Technical Specification Group Core Network and Terminals; Numbering, addressing and identification. 3GPP, December 2012.

[7] 3GPP. TS 23.203 Policy and charging control architecture. 3GPP, December 2012.

[8] 3GPP. TS 33.102 3G security; Security architecture. 3GPP, December 2012.

[9] 3GPP. TS 33.210 3G security; Network Domain Security (NDS); IP network layer security. 3GPP, December 2012.

[10] 3GPP. TS 33.401 3GPP System Architecture Evolution (SAE); Security architecture. 3GPP, December 2012.

[11] 3GPP. TS 43.020 Security related network functions. 3GPP, December 2012.

[12] D. Dolev and A. Yao. On the security of public key protocols. IEEE Transactions on Information Theory, 29(2): 198—208, March 1983.

[13] M. Blaze, W. Diffe, R. Rivest, B. Schneier, T. Shimomura, E. Thompson, and M. Wiener. Minimal key lengths for symmetric ciphers to provide adequate commercial security. Report of Ad Hoc Panel of Cryptographers and Computer Scientists, January 1996. Available from http://www.crypto.com/papers/.

[14] N. Smart (Ed.). ECRYPT II Yearly Report on Algorithms and Keysizes (2010–2011) ECRYPT II NoE, ICT-2007-216676, Deliverable D.SPA.17, Rev.1, June 2011.

[15] R. Anderson. Why cryptosystems fail. In Proceedings of the 1st ACM Conference on Computer and Communications Security (CCS'93). ACM Press, 1993.

[16] B. Schneier. Beyond Fear: Thinking Sensibly about Security in an Uncertain World. Springer, 2003.

[17] D. Florêncio and C. Herley. Where do all the attacks go? Microsoft Research, Technical Report 2011-74, 2011. Available from http://research.microsoft.com/pubs/149885/WhereDoAllTheAttacksGo.pdf.

[18] D. Pavlovic. Gaming security by obscurity. In Proceedings of the 2011 Workshop on New Security Paradigms Workshop (NSPW 2011), 2011.

[19] The rising cost of identity theft for consumer. In Bucks Blog, New York Times, 2011/02/09, 2011.

[20] Fighting fraud together; A strategic plan to reduce fraud, 12 October 2011. Home Office, UK, 2011.

[21] S. Kent and K. Seo. RFC 4301: Security architecture for the Internet protocol. IETF RFC 4301, December 2005.

[22] G. M. Køien and V. A. Oleshchuk. Spatio-temporal exposure control: An investigation of spatial home control and location privacy preserving issues. In Proceedings of the 14th IEEE International Symposium on Personal, Indoor and Mobile Radio Communications (PIMRC 2003), Beijing, China, 7–10 September, pp 2760–2764. IEEE Press, 2003.

[23] D. Erickson et al. A demonstration of virtual machine mobility in an OpenFlow Network. In Proceedings of ACM SIGCOMM'08, 17–22 August 2008.

[24] E. Aïmeur and D. Schönfeld. The ultimate invasion of privacy: Identity theft. In Proceedings of the Ninth Annual International Conference on Privacy, Security and Trust (PST11), Monteral, Canada, August 2011.

[25] Office of the Information and Privacy Commissioner of Ontario. In Privacy by Design: Time to Take Control. www.privacybydesign.ca, Ontario, Canada, January 2011.

[26] G. M. Køien. Privacy enhanced device access. In Proceedings of MobiSec 2011, Aalborg, Denmark, May 2011.

[27] G. M. Køien. Privacy enhanced cellular access security. In Proceedings of the 2005 ACM Workshop on Wireless Security, pp. 57–66, Cologne, Germany, September 2005.

[28] G. M. Køien and V. A. Oleshchuk. Location privacy for cellular systems; Analysis and solution. PET 2005, Cavtat, Croatia, LNCS, Vol. 3856. Springer, 2005.

[29] V. A. Oleshchuk and G. M. Køien. Security and privacy in the cloud; A long-term view. In Proceedings of Wireless VITAE, pp. 1–5, 2011.

[30] ISO, ISO/IEC 11990-1 Information technology – Trusted Platform Module – Part 1: Overview, 2009.

[31] N. Ulltveit-Moe, V. A. Oleshchuk, and G. M. Køien. Location-aware mobile intrusion detection with enhanced privacy in a 5G context. Wireless Personal Communincations, 57(3), 2010.

[32] A. Schwartz. Privacy and security identity management and privacy: A rare opportunity to get it right. Communications of the ACM, 54(6): 22–25, June 2011.

[33] C. Boyd and A. Mathuria. Protocols for Authentication and Key Establishment. Springer Verlag, 2003.

[34] L. Rajbhandari and E. Snekkenes. Intended actions: Risk is conflicting incentives. In Proceedings of the 15th International Information Security Conference (ISC 2012), LNCS, Vol. 7483, pp. 370–386. Springer, 2012.

Biography

Geir M. Køien is an associate professor at the University of Agder, Norway. His primary research interests are system security, personal privacy and cellular access secuirty. He has previously worked for Telenor R&D, where he was a delegate to the 3GPP SA3 security work group for 10 years. Currently he also holds an adjunct position with the Norwegian Post and Telecommunications Authority as a senior advisor on cellular security.

Author Index, Volume 1 (2012)

Keyword Index, Volume 1 (2012)

Online Manuscript Submission

The link for submission is: www.riverpublishers.com/journal

Authors and reviewers can easily set up an account and log in to submit or review papers.

Submission formats for manuscripts: LaTeX, Word, WordPerfect, RTF, TXT.
Submission formats for figures: EPS, TIFF, GIF, JPEG, PPT and Postscript.

LaTeX

For submission in LaTeX, River Publishers has developed a River stylefile, which can be downloaded from http://riverpublishers.com/river_publishers/authors.php

Guidelines for Manuscripts

Please use the Authors' Guidelines for the preparation of manuscripts, which can be downloaded from http://riverpublishers.com/river_publishers/authors.php

In case of difficulties while submitting or other inquiries, please get in touch with us by clicking CONTACT on the journal's site or sending an e-mail to: info@riverpublishers.com

www.ingramcontent.com/pod-product-compliance
Lightning Source LLC
Chambersburg PA
CBHW061828220326
41599CB00027B/5219